Analyzing Media

REVISIONING RHETORIC
Karlyn Kohrs Campbell and Celeste Condit, *Series Editors*

Analyzing Media: Communication Technologies as Symbolic and Cognitive Systems
James W. Chesebro and Dale W. Bertelsen

ANALYZING MEDIA

Communication
Technologies
as Symbolic
and Cognitive
Systems

JAMES W. CHESEBRO
DALE A. BERTELSEN

THE GUILFORD PRESS
New York London

©1996 The Guilford Press
A Division of Guilford Publications, Inc.
72 Spring Street, New York, NY 10012

Printed in the United States of America

This book is printed on acid-free paper.

Last digit is print number: 9 8 7 6 5 4 3 2 1

Library of Congress Cataloging-in-Publication Data
Chesebro, James W.
    Analyzing media : communication technologies as symbolic
and cognitive systems / by James W. Chesebro and Dale A. Bertelsen.
        p.   cm. — (Revisioning rhetoric)
    Includes bibliographical references and index.
    ISBN 1-57230-154-6
    1. Mass media criticism. 2. Communication and technology.
3. Communication and culture.   I. Bertelsen, Dale A.
II. Title   III. Series
P96.C76C48   1996
302.23—dc20   96-29009
                                                    CIP

Preface

For the last 25 years, we have been particularly concerned with what critics of communication have and have not been doing. As a point of departure, we would underscore two of these concerns.

Critics have focused on the content and form of verbal and nonverbal presentations. Indeed, they have tended to treat the content and form of a presentation as the totality of its message. In this regard, communication critics have traditionally failed to give the channel, format, and medium of communication the same attention they have devoted to the ideational presentation of a message.

Furthermore, critics have acted as if the study of certain behavioral responses to media systems is the domain of and more appropriately examined by communication researchers committed to the use of social scientific methodologies. Hence, communication critics have largely bypassed dealing with the behavioral implications of topics such as television programs designed for children, the meaning of pornography, or the social significance of commercial advertising in the television industry.

When it comes to a direct treatment of communication media, again, the content of the media has received the predominance of attention. Hence, the ideology expressed by specific politicians, the sexual and violent content of television and films, and the persuasive efforts of advertisers have been the preferred object of study for most communication critics. Indeed, the issues involved in how messages are produced or formatted in each medium have been viewed as technical, rather than critical, ones. Communication critics might readily agree that technologies create messages, but, at the same time, they will bypass any responsibility for dealing with computer-mediated communication. Accordingly, for the communication critic, the impact of the choice of a medium and the values conveyed by one medium rather than another have been slighted.

However, we do not mean to isolate only communication critics as the audience that motivates our analysis in this volume. We are convinced that most people, like communication critics, have also ignored and thereby bypassed responsibility for responding to the format, behavioral consequences, and message-generating power of technologies.

In this volume, we hope to offer an alternative way of responding to communication technologies. Certainly, we think it is essential that people respond to the content and form of the ideas explicitly conveyed through the communication media, but we think it is equally important that each of us also be conscious that the technology of a communication system can, of and by itself, generate and convey messages in addition to the messages conveyed by the ideas transmitted through a communication medium. The alternative approach we outline in this volume will allow each of us to respond more critically to the messages and values conveyed by communication technologies, regardless of the particular content and form of the ideas intentionally conveyed through a communication technology.

In other words, we think there are at least two kinds of messages conveyed by every communication technology. One message is conveyed in consciously and intentionally articulated ideas presented by human communicators or persuaders, although sometimes these ideas can be more implicit than we realize and the motives for them even unclear or unknown to us. Traditionally, communication critics have directed our attention to the importance of these ideas. This emphasis is important if each of us is to survive in our symbolic world. However, a second message is also conveyed by the communication technology or format selected to present an idea. We hope to demonstrate that each of us must pay attention to the message that these communication technologies convey. This volume identifies what these communication technology messages are and how we can more adequately prepare ourselves to recognize and respond thoughtfully and critically to them.

This volume is divided into four parts.

Part I, "Introduction," is made up of two chapters that set the context for our approach. In Chapter 1, "A History of Human Communication," the emerging role of communication technologies is highlighted and the communication technologies that become our objects of study are described. In Chapter 2, "A World of Communication Technologies and the Human Response," we provide a survey of how communication researchers and critics have responded to communication technologies. The four viewpoints considered are the "Media Consumption," "Uses-Gratification," "Cultivation," and "Critical" approaches.

Part II, "The Critical Moment and the Critic's Method," sketches an approach to communication that emphasizes the format of communication technologies. The production systems that create communication technologies and account for the messages of these communication technologies are a significant, but often neglected, foundation for critics' analyses. Chapter 3, "The Critical Process," outlines our approach to criticism, especially to some of the problems that emerge when defining media criticism and the basic principles required to assess communication technologies. These considerations provide a foundation for describing the characteristics of criticism and for outlining a method for describing, interpreting, and evaluating technocultural dramas.

In Part III, "Media Cultures," we consider the influence of specific communication technologies on social organization and human understanding. Chapters 4–6 are applications of the conceptions outlined in the previous chapter. Chapter 4, "The Oral Culture," examines cultural systems dominated by face-to-face verbal and nonverbal communication systems. Chapter 5, "The Literate Culture," describes cultures governed by writing and print communication systems. Chapter 6, "The Electronic Culture," examines how televised and computerized communication is altering how and what people understand.

Part IV, "A Future Perspective," explores potentialities. Chapter 7, "Analyzing Media Comparatively: Comparative Media Criticism and the Future of Media Criticism," provides a formal comparison of selective features of each communication technology along precise and narrowly defined criteria. The comparisons provide a context for suggesting future directions for the analysis of communication technologies.

In addition, we include a Glossary that explores five key concepts used throughout this book: *communication, cognition, culture, technologies,* and *technologies as communication systems.* For students of media criticism, the Glossary provides a sense of the complex human processes underlying these familiar, but hard to define, terms. For media critics as well as students, the Glossary entries explicitly reveal our perspective on the objects of our study. Readers are directed to the Glossary entries at appropriate points in the text.

In all, each of us now lives in environments regulated, if not controlled, by communication technologies. The content of these communication technologies has attracted attention and been the object of critical observations. Yet, the content of these communication systems can change daily. The presence and functions of the communication systems themselves remain. We are proposing that we examine how these communication technologies are affecting the quality of human life. We suspect, for example, that communication technologies constitute the organizing principle determining human patterns and habits. It is easy to imagine a "typical" 24-hour scenario in which an individual wakes up to a clock radio program, keeps a morning television show on during the breakfast hour "solely" as "background," drives to work with the car radio on, interacts all day long with a host of diverse communication technologies at work, drives home with the car radio on, and finally, later that night at home, watches some 3 hours of television until retiring. While we might carefully examine the content of each of these contacts with communication technologies, we might also pay equal attention to the fact that a person's entire day is a series of contacts with one kind of communication technology after another. And, during "special" events, such as political campaigns and elections or national disasters, communication technologies become even more important in our lives. In our view, we need to become critically aware of how these systems are affecting what we see, what we select to organize in our lives, how we organize our lives, what we know, how we know things, and perhaps even why we know what we know. Both the content of communication technologies and the communication technologies themselves must be assessed when we seek to respond effectively, thoughtfully, and creatively to the quality of our lives. We hope that this volume makes a contribution toward this end.

Acknowledgments

Several colleagues were particularly helpful with this project. Our words were made more precise and thoughtful with the kind and insightful feedback of Celeste Condit, Karlyn Kohrs Campbell, and Peter Wissoker. We dedicate this book to Donald G. Bonsall and Mary Mino, "without whom not," to borrow a dedicatory phrase from Kenneth Burke, our friend and former colleague. Don and Mary, we completely appreciate all of your continuing support in this endeavor and all of your assistance and corroboration in the other projects we have undertaken in past years. Thank you!

Contents

PART I Introduction

P art I, "Introduction," is composed of two chapters. The order of these chapters should be
viewed with a great deal of flexibility. Either
chapter may be read first, depending upon the background and interests
of the reader.

Chapter 1, "A History of Human Communication," offers a view of
the history of communication that is chronological and developmental in
its organization, seeking to explain the evolution of communication by
identifying and highlighting critical and decisive moments in its annals.
However, this history is unusual in that it does not, for example, focus on
outstanding human communicators, nor on social movements and campaigns that have shaped human interactions. Instead, it accounts for
changes in human communication by centering on the means or technologies that people have used to communicate. In particular, this chapter
draws attention to changes in communication technologies, and it seeks
to mark how these technological changes have affected how humans
communicate. In this view, the invention of the printing press is an
outstanding moment in the history of communication, and the patterns
of human communication before and after this innovation are compared.

In Chapter 2, "A World of Communication Technologies and the
Human Response," we provide a survey of how communication researchers and critics have responded to changes and developments in
communication technologies. We specifically focus on how communication technologies have been studied. The four viewpoints considered are
the "Media Consumption," "Uses-Gratification," "Cultivation," and
"Critical" approaches.

1

A History
of Human
Communication

Human communication is an ever-changing process, and criticism is one response to its evolution. Critics are influenced consciously and unconsciously by the history of human communication. Indeed, critics frequently participate in constructing this history. DeFleur and Ball-Rokeach (1989), for example, have noted that "previous revolutionary changes in the capacity of people to share meanings with others have had truly powerful influences on the development of thought, behavior, and culture" (p. 4). They concluded, "Understanding these changes and their consequences will make it easier to appreciate an important aspect of our contemporary mass media: Even though they have only recently arrived, they are already so central to our daily lives that they may help shape the destiny of our species in the future" (p. 4). Such histories are persuasive ideological constructions that reconceive the past, present, and future, and that govern assumptions about the history of human communication (see, e.g., Andrews, 1968; Blair & Kahl, 1990; Noble, 1965).

In this chapter, we suggest how human beings evolved into the kinds of communicators they are. We hope this historical sketch offers a useful view of significant moments in the history of human communication as it reveals the understanding of criticism that dominates this book. We have several objectives herein. Initially, prior approaches to understanding the history of human communication are identified. They establish a context for considering an approach that emphasizes the role of technology and culture in explaining the evolution of human communication. From this perspective we offer a brief history of human communication, best understood as a synoptic overview. We conclude with an explicit discussion of the principles underlying this history. These principles reveal several of the essential features of the paradigm of criticism used in this volume and, thus, introduce the treatment of criticism found in later chapters and in the discussion of key terms that constitute the Glossary at the end of this volume.

| PRIOR HISTORICAL APPROACHES

Previous approaches have varied in suggesting how human *communication* (see Glossary) developed into the system we now have. Each of these approaches is particularly useful in understanding particular features of the evolution of human communication, and each warrants recognition for the contribution it has made to our understanding of human communication.

Some have emphasized the emergence of symbol using, such as Burke's conception of the evolution of human communication from signal to symbolic communication. In the early 1950s, Burke (1952a, 1952b, 1953a, 1953b) highlighted the emergence of the word "no" as a nonreferential concept that illustrates, in his view, the "pure" use of the symbolic.

The history of the development of rhetoric has been a dominant focus in tracing the history of human communication. Ehninger's (1968) essay outlining three eras in the development of Western rhetorical thought often serves as the traditional approach to this practice. He has argued that the development of rhetorical thought could be divided into a classical period wherein the grammar of the speech act was privileged, followed by a period through the late 18th century wherein the speaker–listener relationship was privileged, culminating in a period from the 1930s to the late 1960s wherein sociological concerns were of upmost importance. More contemporary approaches are diverse and widespread, covering thousands of years and a host of different rhetorical trends or tendencies (see, e.g., Blair & Kahl, 1990; Brinton, 1990; Makus, 1990; J. Poulakos, 1990; T. Poulakos, 1990).

Others have focused on the development of theories of rhetoric that have emerged from the study of public address as an approach to the evolution of human communication. Thonssen, Baird, and Braden (1970) examined the development of rhetorical theory and identified the "foundations of the art of speaking," the "extensions of basic principles," and the "contributions of the modern theorists" (p. 33). Rather than positing dramatic changes in the evolution of human communication, they offered a conception of human communication that emphasized its coherence and unity. In their view, "rhetoric derives from an extensive literature" that "helps to give continuity to our efforts" and creates "a common core of theory" (Thonssen et al., 1970, p. 33). The volumes produced by Brigance (1943/1960), Clarke (1953), Hochmuth (1955), Clark (1957), Baldwin (1959), Auer (1963), Boase (1980), Ryan (1983), and Johnson (1991) all suggest how insightful and comprehensive such an approach can be.

Still others have written histories of specific social groups, such as Campbell's (1986, 1989a, 1989b) analysis of the rise of women speakers (see also Vonnegut, 1992), A. Smith and Robb's (1971) historical treatment of African American discourse (see also Scott & Brockriede, 1969), and Darsey's (1981, 1991) study of the evolution of gay male and lesbian communication (see also Ringer, 1994).

Developments within the discipline of communication have been the focus of other historical approaches. The emergence of specific methods for the study of human communication, such as Delia's (1987) history of the emergence of social scientific methods generally (see also Marvin, 1985; McQuail, 1980) or in specific content areas (see, e.g., Chaffee & Hochheimer, 1985; Sloan, 1991; Weaver & Gray, 1980) has received attention. Benson (1985b; 1989, especially pp. 28–40, 90–108, 130–156, 196–220; 1992) has made more comprehensive surveys of the evolution of theories and methods in several specific areas of human communication (see also Bormann, 1965, pp. 16–29; Phillips & Wood, 1990). Others have focused on the development of professional associations (e.g., H. Cohen, 1985, 1994; Lutz, Huber, Arnold, Wilson, & Chesebro, 1985; Oliver & Bauer, 1959; Rogers, 1994; Wallace, 1954; Work & Jeffrey, 1989). Finally, using specific figures within the discipline, some theorists have fashioned conceptions of the discipline from the perspective of its founders (see, e.g., Reid, 1978, 1981, 1990; Speech Communication Association, 1964; Windt, 1990).

All of these histories of human communication are useful depending on the kinds of questions being raised. We are, however, particularly interested in the emergence of communication technologies, how technologies interact with cultural systems, and the influence of technologies and cultural systems on human communication. Accordingly, a cultural and technological view of the evolution of human communication is central to our concerns here.

A CULTURAL AND TECHNOLOGICAL PERSPECTIVE

Technology and cultural systems have been recognized as variables affecting the evolution of human communication, but frequently they are treated as separate and independent variables. Adopting a technological orientation, some theorists have traced the emergence of technologies as governing determinants of human communication systems (see, e.g., Stevens & Garcia, 1980; DeFleur & Ball-Rokeach 1989, pp. 3–45, especially pp. 46–122). Others have accentuated cultural influences on human communication (see, e.g., Bremmer & Roodenburg, 1991).

The interaction between *culture* and *technology* (see glossary) is a particularly rich and rewarding study, and a host of scholarly endeavors are the foundation for our view of the history of human communication. We are profoundly influenced by some of these scholars, including Auerbach (1953), Havelock (1963), Clanchy (1979), Ong (1982), Logan (1986), and Lentz (1989). Although each examined a specific issue, many of their observations are historical, and their analyses are particularly relevant in constructing a history of human communication. We have integrated their understandings into a single historical framework that we believe has dominated the evolution of human communication.

This cultural and technological perspective of the history of human communication raises several important questions. These include the following: Has technology influenced human communication? Have technologies actually determined how systems of human communication have developed? Do cultural systems affect how communication technologies are used? How have technology and culture affected human understanding and cognition?

We respond to these questions by first proposing a scenario describing the history of human communication, followed by a discussion of the principles that forged this history.

A HISTORY OF HUMAN COMMUNICATION: TECHNOLOGICAL AND CULTURAL INFLUENCES

Three major stages or benchmarks in the history of human communication are noteworthy. These are the creation of the oral, literate, and electronic cultures.

The Oral Culture

The oral culture can be traced back to the beginning of *Homo sapiens* some four to five million years ago, but oral languages as we now understand them emerged 4,000–5,500 years ago. The growth of the global communication system, fostered by the mass production and distribution of books as well as by electronic media, has led to the view that oral modes of communication are not as important as they once were. Yet for the vast majority of world cultures, orality continues to be the primary means by which societal values are transmitted from one generation to the next. Ong (1982, p. 7) has reported that of all of the thousands of languages, possibly tens of thousands, spoken in the course of human history, only around 106 have produced literature. Similarly, Edmonson (1971) concluded that only 3 percent of the spoken languages in the world today have a literature, and Maheau (1972, p. 83) has reported that two-fifths of the adult population of the globe—more than 700 million people—cannot read or write. Applying the United Nations Educational, Scientific, and Cultural Organization's (UNESCO's) definition of the minimum requirements for mass communication systems to 169 world cultures, Chesebro (1985b, 1986b) has reported that 51 percent of the world's nation-states are "pre-mass oral cultures" or "mass oral cultures," in contrast to the 11 percent that are "mass literate cultures," the 36 percent that are "mass literate-electronic cultures," and the 3 percent that are predominantly "mass electronic cultures." McHale (1972, p. 83) has reported that two billion people throughout the world do not have the basic facilities to use television, radio, and film as mass media. In all, orality is the world's dominant mode of communication. For the vast majority of the world's cultures, orality is the system that creates and sustains cultural values, concretizes the unique understandings embedded within those values, and transmits these values from one generation to the next. Based on these data, we use *oral culture* rather than oral communication because it more fully captures the dimensions of the object we are investigating. As we use the term, oral communication is not merely a mode of or channel for conveying information, but instead the primary perceptual and cognitive framework by which values are traditionally created, defined, sustained, and transmitted from one generation to the next.

At the same time, an oral culture is a product of a long and complicated process rather than of a single moment in which human beings began to communicate orally. Buettner-Janusch and Day (1987), professors of anthropology and anatomy, respectively, have reported that "no evidence exists to show how hominid language first developed," but they have also noted that "some generalizations from modern man may be justified" (p. 947). In this context, they have reported that the emergence of oral communication in early human beings depended on the development of "the cortex of the brain, an area that has expanded rapidly in hominid evolution" (p. 946) and the downward shift of the larynx "directly reflecting development of erect posture and expansion of the braincase" (p. 947). Such evolutionary changes, marked by fossil and toolmaking remains, provide a context for identifying specific stages in the development of oral communication.

Signal Communication

The first efforts to communicate were predominantly by nonlinguistic gestures and body movements. Vowel sounds later emerged, we would suspect, to emphasize these gestures. With time, neuroanatomical developments in the human being allowed gestures and vowel sounds to function as equally important dimensions in human interactions. As Buettner-Janusch and Day (1987) have suggested, "vowel sounds had their origins in nonlinguistic vocalizations" and "consonants were added as the hominids developed more control over their airways by manipulating tongues, lips, and teeth" (p. 947). At this stage in human development, gestures, body movements, and oral sounds would necessarily be linked to the representation of physical phenomena. Under such circumstances, gestures, body movements, and oral sounds would be externally oriented and function as devices designed to identify circumstantial events and relationships.

At the same time, these devices—gestures, body movements, and oral sounds—introduced new experiences into the life of the human being, for they created a new class or type of sensory data that had not previously existed. As these devices became a part of the learned experiences of human beings, they began to create a new system, aptly identified as *signal communication*. Signal communication led to profound, neuroanatomical changes in the human beings who used it, and also created a new awareness of the uniqueness of others as well as new social networks based on its use. As Buettner-Janusch and Day (1987) have reported, "Coupled with the ability to make the sounds necessary for speech would be changes in the brain that allow vocabulary to be stored and retrieved and changes in the auditory apparatus that allow language to be understood when spoken by others with slightly different intonation or pitch" (p. 947).

Without precise evidence, the functions of signal communication at this early stage in human development can only be hypothesized. However, the signal communication system of early human beings should have allowed them to achieve five objectives, as described by Ekman and Friesen (1969), who have provided a repertoire of nonverbal behaviors that are consistent with what is known about the development of early human beings. The authors have reported that nonverbal communication can function as emblems (i.e., when nonverbal acts function as equivalents of verbalizations), or to illustrate (i.e., visualize clarifiers and explainers), to display emotions (i.e., provide affective "statements"), to regulate or control, and, finally, to adapt to the self, others, and the environment.

At the same time, we have no evidence for suggesting that the functions that might have been achieved with the signal communication systems of early human beings can or should be understood as a product of the motivational schemes now associated with contemporary human beings. The possibility of committing an intentional fallacy looms extremely high. Indeed, Jaynes (1976) has suggested that the origin of consciousness may have been a product of the emergence of the bicameral brain. He has suggested that the communication of early human beings from as early as 3000 B.C. to as late as 1230 B.C. may have emanated from "a mentality utterly different from our own" (p. 68), and he has specifically argued that the "beginnings of action" were not to be found "in conscious plans, reasons, and motives" (p. 72). Similarly, Gordon (1971) has identified a solely signal period in the evolution of human communication that he characterizes as an "epoch of prehistory" (p. 26). Likewise, particularly when examining human evolution as a system, Bertalanffy (1968) has argued that a solely signal period of communication existed when human beings were necessarily motivated by only "biological drives" (pp. 216–219). He has suggested that "creative proception in contrast to passive perception," the ability to objectify "both things outside and the self," and "intention as conscious planning" must be examined within an "evolutionary framework" (p. 216). Distinguishing signal and symbolic communication systems, Bertalanffy has specifically argued that "the root of creative symbolic universes cannot therefore be reduced to biological drives, psychoanalytic instincts, reinforcement of gratifications, or other biological factors" (p. 216). These perspectives suggest that it is appropriate to distinguish signal and symbolic communication and to recognize that the communication system of early human beings may have been predominantly a signal system, and that a shift from a signal to symbolic orientation may parallel a series of related human developments, such as the emergence of human languages.

Symbolic Communication

The shift from *signal* to *symbolic* communication constituted a fundamental change. Before identifying how this shift affected human civilization, it is appropriate to distinguish signal and symbolic communication. Cassirer (1944/1965) has provided an initial base for distinguishing these two modes of communication. He initially distinguishes symbols and signals: "Symbols—in the proper sense of this term—cannot be reduced to mere signals. Signals and symbols belong to two different universes of discourse: a signal is part of the physical world of being; a symbol is a part of the human world of meaning" (p. 32). In Cassirer's view, "Signals are 'operators'; symbols are 'designators' " (p. 32). Thus, Cassirer concludes, "Signals, even when understood and used as such, have nevertheless a sort of physical or substantial being; symbols have only a functional value" (p. 32).

Signs and symbols both function as forms of communication when two or more people assign the same meaning to them. However, a sign functions as signal communication when the sign creates a shared universe of understanding of "the objective reference of a word. . . . For example, the denotative meaning of 'pencil' is *that which writes*. There are no personal interpretations of this meaning; it states an objective fact" (Blankenship, 1968, p. 20). In contrast, a symbol functions as symbolic communication when the symbol creates a shared universe of understanding

> beyond the objective meaning of the words. In fact, a word may have a purely objective meaning for one listener and a highly colored connotative meaning for another. Consider the word "spider." To the scientist accustomed to dealing impersonally with spiders, the word "spider" equals arachnid. But to the small child who has been badly frightened by a spider, the word "spider" not only carries with it the child's equivalent of "arachnid" but arouses a fear response as well. (Blankenship, 1968, p. 21)

In all, signal communication includes the use of words and nonverbal behaviors that create a shared universe of understanding regarding the objective referent of a word or nonverbal behavior; signs are denotative and predominantly identify the physical existence, physical characteristics, and physical functions of external phenomena. In contrast, symbolic communication includes the use of words and nonverbal behaviors that create a shared universe of understanding regarding the subjective associations of a word or nonverbal behavior; symbols are connotative, socially constructed by human beings, and emphasize the values people attribute to words and nonverbal behaviors.

The moment at which human beings began to use words and nonverbal behaviors both as signals *and* symbols may well have been the birth of civilization and the *Homo sapiens* (human beings as critical thinkers). Cassirer (1944/1965) has noted, for example, that there was a moment at which human beings "developed a new form: a *symbolic imagination and intelligence*" (p. 33). Burke (1952a) has argued that this new form of symbolic communication emerged when human beings used the word "no," for the "negative is a peculiarly linguistic resource" that does not exist in nature in any form (p. 251). As symbolic communication increased, human beings began to develop as we know them today, or as Walter and Scott (1973) have argued, "The growth and development of symbolization is almost synonymous with human growth and development" (pp. 240–245).

In human beings, the development of symbolic communication was marked by neuroanatomical changes. Buettner-Janusch and Day (1987) have noted that "the complexity of human behaviour is related to the human ability to interpret symbols, to appreciate abstract ideas, and to communicate them to others, particularly the young. Neuroanatomical studies show that these abilities reside primarily in the cortex of the brain, in areas that have expanded rapidly in hominid evolution" (p. 946). In greater detail, they have explained that the

> volume alone is not enough, and the level of differentiation and organization of brain tissue may also be of critical importance. The fossil record can yield only endocranial casts and, from them, possible brain volumes, but the firm association of stone tools with such remains must indicate a level of intellectual attainment that can foresee a use for a tool, envisage it within a stone, and then shape it to a set and repeatable pattern. (p. 946)

The ability to symbolize allowed human beings to create languages. In this context, a language is a social system of conventional and arbitrarily spoken symbols. When a language is defined in this fashion, special attention should be devoted to the fact that a language is social (a human creation or "contract" and agreement with others), systematic (i.e., an organized and established pattern creating redundancy and therefore predictability), conventional (i.e, a set of rules of conduct and behavior), and arbitrary (i.e., the creation and selection of specific words may have no substantial relationship to physical objects).

It is unknown when the first language was created. Williams (1982) has maintained that *Homo sapiens* developed the first language system in 34,000 B.C. (p. 28). Reflecting a larger consensus of opinion, the creation of the first language probably occurred around 5000 B.C. DeFleur and Ball-Rokeach (1989) have noted that modern linguistics "have identified large numbers of words in some fifty prehistoric vocabularies and in numerous modern languages that can be traced back as far as about 5,000 B.C. (some 7,000 years ago) to a proto-Indo-European 'common source'" (p. 17). They have concluded that "it can be suggested that this common source ultimately led back to the language originally developed by the Cro Magnon people. In any case, there is no question that the development of speech and language made possible great surges in human development" (p. 17). In this context, Buettner-Janusch and Day (1987) have specifically identified the new possibilities created when human beings began using languages. They have maintained that spoken language "allows human beings to name things with 'open' symbols; i.e., symbols that, in countless combinations, can be made to relay different messages. Not only can human communication cover immediate situations and feelings but also discussions at abstract or hypothetical levels" (p. 947). As a result, concluded Buettner-Janusch and Day, "the human thus can store and transmit knowledge gained by past experiences, as well as discuss plans for the future" (p. 947).

The use of oral languages dramatically altered human behavior, particularly the ways in which human beings interacted with each other and organized themselves. In Chapter 4, in greater detail, we characterize the essential features of oral language as a human system as well as identify and examine the new interactional patterns created by oral languages. At this point, we would observe that oral languages created a new cultural system.

Because we now live in a social system mediated by oral, literate, and electronic modes of communication, we lack references that would allow us to experience and understand a culture governed solely by oral languages. Haynes (1988) has provided a hypothetical example that begins to reveal the cultural system created and sustained by a solely oral system of communication:

> One promising avenue is to consider what a nonliterate classroom might be like. No text, of course; no lecturing from outlines or notes, and, certainly, no notes would be taken. Oral discourse of a totally practical nature, fully interactive, conducted by a teacher who maintains active intellectual engagement with the audience would be the norm. Such a class might cover a body of theoretical material developed through literate means, but if the end is a practical one, that material would be considered only in the context of actual experience. As the class is led to experience, for example, changes in their predispositions, stimulated either through their own behavior or through that of the teacher, such experience is framed in theoretical terms, always with direct reference to immediate events. We have in mind, by the way, skills-oriented courses in human communication. To gauge the extent one's own institution is biased by literacy in the sense discussed above, one need only imagine the furor such a nonwriting based course would create. (pp. 96–97)

As Haynes's example suggests, a culture mediated by only oral communication would necessarily differ dramatically from cultural systems dominated by writing, print, and electronic modes. Ong (1982, pp. 37–57) has identified several distinct features of a culture mediated by orality that we examine in Chapter 4. At this juncture we need only illustrate how systems of thought change as one moves from the oral to literate to electronic culture. The nature of knowledge functions as a useful example, and, as we might anticipate, the nature of knowledge changes as one moves from an oral to literate to electronic culture. Explaining these shifts, Chesebro (1989) has noted,

In an oral culture, the knower and what is known are related. Accurate and reliable knowledge require direct social interaction, participation in the lived experience, and exposure to the imminent and immediate source of knowledge. However, in a literate culture, the knower and what is known are typically unrelated. The sociological and personal features of the source of the printed word are unlikely to be known. Indeed, in a literate culture, sources are likely to be, at best, ambiguous and receivers unpredictable. Once our words appear in print, we have no idea who will read them or how they will react. In contrast, the nature of knowledge is dramatically different in electronic modes such as television and film. While the knower and what is known are reunited in electronic media such as television and film, knowledge is separate from the lived experience. Television and film conceptions of "what is" report only what can be seen and heard; but more importantly, this visual and auditory conception of "what exists" is typically understood within a totally unrelated context (the home or the movie theater), a context which did not characterize the original situation. Context-defining influences are thus lost in the electronic culture. (p. 12)

As these examples suggest, each medium of communication creates a selective perception of reality (Chesebro, 1984, p. 116–124). The oral mode draws attention to the immediate presence of the speaker, the physical relationship between speaker and listener, and the speaker's unique verbal and nonverbal behaviors, all of which are understood within the immediate context in which the interaction occurs. Accordingly, perception or what is "understood to be" in the oral context is a function of the appreciation and experience of the merger of or the forging of linkages among the immediate elements in the communication process. As Ong (1977) has noted, "Oral utterance thus encourages a sense of continuity with life, a sense of participation, because it is itself participatory" (p. 21)

Additionally, without the benefit of writing, print, and electronic recording devices, if ideas are to be retained, they must be memorized. But, as Opland (1983, p. 158) has noted, even if mnemonic devices are used and verbatim recall is sought, repetition produces agreement with the original source only 60 percent of the time. What is permanent and what can be permanent within an oral culture is thus dramatically different than in literate and electronic cultures, simply because writing, print, and electronic recording devices virtually eliminate recall as an issue. If the oral culture is to sustain itself, if the values of the community are to be transmitted from one generation to the next, then delivery and memory must necessarily become its critical rhetorical canons. As we shall suggest later, given their technological constraints, the canons of style, arrangement, and invention become far more important in cultural systems mediated by literate and electronic communication systems (Chesebro, 1989, pp. 9–17).

The Literate Culture

Webster's Third New International Dictionary of the English Language Unabridged and Seven Language Dictionary (1986) defines a *literate* person as "one who can read and write" and *literacy* as "an ability to read a short simple passage and answer questions about it" (p. 1321). The mastery of two modes of communication, *writing* and *print*, established literacy as a critical human skill and ultimately became a standard for defining what civilized and civilization mean. Noting that "literacy is in no way necessary for the maintenance of linguistic structure or vocabulary," Robins (1987, p. 585), a professor of general linguistics, has argued that "literacy has many effects on the uses to which language may be put," including the "storage, retrieval, and dissemination of information [which] are greatly facilitated [by literacy], and some uses of language, such as philosophical system building and the keeping of detailed historical records, would scarcely be possible in a totally illiterate community. In these respects the lexical content of a language is affected, for example, by the creation of sets of technical terms for philosophical writing and debate." Robins has concluded, "Because the permanence of writing overcomes the limitations of auditory memory span imposed on speech, sentences of greater length can easily occur in writing, especially in types of written language that are not normally read aloud and that do not directly represent what would be spoken" (p. 585). In this regard, the development of two modes of communication, *writing* and *print*, has fostered the unique functions of literacy, which are considered here one at a time.

Writing

The first oral language, formulated in approximately 5000 B.C., some 7,000 years ago, preceded by some 3,000 years the development by the Seirites of the first principles of phonetic alphabetic writing—known as the Proto-Sinaitic script—in the Sinai and Canaan in 2000 B.C. The Seirites, referred to in the Bible as the Kenites or the Midianites, sojourned with Moses in the desert of Sinai during the exile from Egypt. Some believe that the Seirites may have simplified the hieroglyphics of the Egyptian writing system by selecting and developing only the uniconsonantal signs of the latter stages of Egyptian writing. Regardless of its origins, the Proto-Sinaitic script, according to Logan (1986) operated "according to a phonetic acrostic principle whereby each sign represents an object and the sound value of the sign is the same as the first consonant of the name of the object depicted" (p. 34). In greater detail, Logan has reported,

> Words are spelled out phonetically using the names acrostically. For example cat would be spelled by picturing a can, an apple, and a table in sequence. Some of the signs in the Egyptian and Seirite system were similar. The twenty-two Seirite signs, however, had their own Semitic names and sound values. As with the Egyptian uniconsonantal signs, the Proto-Sinaitic alphabet consisted solely of consonants. The vowels and vocalization of speech were not indicated and had to be filled in by the reader. (p. 34)

Although the Seirites were the first to employ the basic principles of phonetic alphabetic writing, because they were unable to consistently reflect vowels in a phonetic alphabetic system, technically they did not develop the first fully phonetic alphabetic writing system. It was the Greeks, some 500 years later, in 1500 B.C., who invented the first fully developed system of phonetic alphabetic writing.

Clearly, the phonetic alphabet was not directly derived from spoken language. Dating back to 30,000 B.C., the first forms of human notation were actually notches on animal bones to record quantities. To reflect qualitative information, this notation system developed into a system of pictographic signs (i.e., pictography, semasiography, ideography, or sub-writing) in which a written visual image or nonverbal picture is intended to identify, represent, and recall objects or beings. It remains unclear, however, if pictographic signs did or could function as an effective mode of human communication. Goldblatt (1987), a professor of Chinese, has reported, for example, that the first Chinese hieroglyphics were "an impediment to education and the spread of literacy, thus reducing the number of readers of literature; for even a rudimentary level of reading and writing requires knowledge of more than 1,000 graphs, together with their pronunciation" (p. 257).

Following the development of pictographic signs, phonographic systems emerged in which written signs had a correspondence to language sounds, but the written signs continued to reflect a visual orientation (e.g., the oral statement "I saw" might be reflected in written form with an image of an eye and a saw). As developed in the Sumerian writing system, pictographic and phonographic signs gradually began to include phonetic concepts. Subsequently, the Sumerians began to employ one written sign for each word (i.e., a logographic system).

The logographic system evolved into a more fully developed phono-graphic system in which each word is represented by its component syllabic sounds, as reflected in the Proto-Sinaitic script developed by the

Seirites in approximately 2000 B.C. Ultimately, the first fully developed phonetic alphabet was developed by the Greeks in approximately 1500 B.C. Gelb (1987), a professor of linguistics, has maintained,

> If the word "alphabet" is understood as writing that expresses the single sounds (i.e., phonemes) of a language, then the first alphabet was formed by the Greeks. Although throughout the 2nd millennium BC several attempts were made to find a way to indicate vowels in syllabaries of the Egyptian-Semitic type, none of them succeeded in developing into a full vocalic system. . . . while the Semites made sporadic use of these indicators, called *matres lectionis*, the Greeks used them systematically after each syllabic sign. (p. 986)

Accordingly, Gelb has reported that "it was therefore the Greeks who, having accepted in full the forms of the West Semitic syllabary, evolved a system of vowel signs that, attached to the syllabic signs, reduced the value of these syllabic signs to simple consonantal signs, and thus for the first time created a full alphabetic system of writing" (p. 986). From that time, however, few changes have occurred in the basic nature of the alphabet. As Gelb has aptly concluded, "In the past 2,800 years the conquests of the alphabet have encompassed the whole of civilization, but during all this period no reforms have taken place in the principles of writing. Hundreds of alphabets throughout the world, different as they may be in outer form, all use the principles established in the Greek writing" (p. 986). However, the power of the communication system developed by the Greeks was not to be fully realized until the development of print.

Print

As far as we know, the first printing press was a product of sixth century Chinese ingenuity, but mass-produced movable type fonts with one alphabet letter per font were probably invented by Gutenberg in the mid-1440s. As editors Unwin and Unwin (1987) have suggested, "Printing seems to have been first invented in China in the 6th century AD in the form of block printing. Other Chinese inventions, including paper (AD 105), were passed on to Europe by the Arabs but not, it seems, printing" (p. 457). As they trace the evolution of printing, Unwin and Unwin have reported, "The invention of printing in Europe is usually attributed to Johannes Gutenberg in Germany about 1440–50, after a period of block printing from about 1400. Gutenberg's achievement was not a single invention but a whole new craft involving movable metal type, ink, paper, and press. In less than 50 years it had been carried over most of Europe, largely by German printers" (p. 457).

Historically, the effect of printing on book production was amazing. In a 50-year period, from 1450 to 1500, the number of books increased in Europe from "scores of thousands" to "more than 9,000,000" (Unwin & Unwin, 1987, p. 462). Consequently, Bramson and Schudson (1987) have noted: "The invention of the printing press in the 15th century had been the beginning of a movement that spread with the growth of literacy and technological innovations in printing and marketing. The circulation of books and pamphlets spread, and literacy levels in Europe advanced markedly" (p. 941).

The impact of print and literacy was equally significant in the United States. The first printing press arrived in the United States in 1683. In 1741, America's first magazine appeared. The "penny press" was initiated in New York City in 1848, and comic books appeared in 1889.

From a broader but equally important historical perspective, the modes of human communication were also undergoing a massive transformation at an ever-increasing rate of change. Some 30,000 years had passed between the first known uses of signal communication by human beings in 34,000 B.C. to the development and use of the first known oral and symbolic languages in 5000 B.C. Some 3,500 years had passed between the development of languages in 5000 B.C. to the development of the first fully phonetic alphabetic writing system by the Greeks in 1500 B.C. Some 3,000 years had passed between the development of the phonetic alphabet in 1500 B.C. and the contemporary printing press in A.D. 1450. Each of the successive "revolutions" in human communication technologies was occurring more quickly.

We believe that each successive communication technology exerted a greater change on the social fabric of human beings. We would hold that with the development of the written system of communication, the phonetic alphabet fundamentally changed the ways in which human beings understand. The basic unit of human communication became the individual word. In the oral culture, the basic unit of human communication was a complex set of interactions that included the character and role of the speaker, the verbal and nonverbal delivery of the speaker, the context in which the speech occurred, and the listener's sense of the immediacy and relevancy of the entire speaking occasion. In the written mode, human contact is made through words constructed from a phonetic alphabet that may bear little resemblance to what is—or can be—directly experienced in everyday life. The symbolic relationship created between writer and reader was contained in words that, once preserved in written form, could exist as meanings independent of either the writer or the reader or the context in which the words were written and read. For the first time in human history, writing allowed human beings to interact through a series of abstractions, the words created by the phonetic alphabet. In like manner, printing exerted an equally profound effect on human communication. Indeed, Logan (1986), as well as several others, has maintained that print created a series of new social institutions, including mass education and mass literacy, as well as influencing the rise of visual thinking and science, the endurance of the Renaissance and the Reformation, individualism, nationalism, and the Industrial Revolution (see also Clanchy, 1979; Illich & Sanders, 1988; Luria, 1976; Ong, 1982).

Once print and the new social institutions created by print were established, print may also have begun to alter oral communication. As Ong (1977) has noted, "Talk, after writing, had to sound literate—and 'literate,' we must remind ourselves, means 'lettered,' or post-oral" (p. 87), for "writing grows out of oral speech, which can never be quite the same after writing is interiorized in the psyche" (p. 339). Ong has concluded that "writing leads verbalization out of the agora into a world of imagined audiences—a fascinating and demanding and exquisitely productive world. Print grows out of writing and transforms the modes and uses of writing and thus also of oral speech and of thought itself" (p. 339). Such claims intrigue us, and we explore the social implications and consequences of writing and print in Chapter 5.

The Electronic Culture

As we noted earlier, once communication systems began to develop, the successive "revolutions" have been engendering massive transformations at an ever-increasing rate of change. Some 3,000 years had passed between the development of the phonetic alphabet in 1500 B.C. and the contemporary printing press in A.D. 1450, but only 400 years passed between the development of the printing press in 1450 and the first electronic modes of communication in 1844. Rogers (1986) has identified the date most frequently associated with the development of electronic modes of communication, reporting that "the first electronic telecommunication occurred on May 24, 1844, when Samuel Morse, inventor of the telegraph, sent the famous message, 'What hath God wrought?' from Baltimore to Washington, D.C." (p. 29). Rogers has also captured the impact of this change: "Until that time, information could travel only as fast as the messengers who carried it; communication of information and transportation of people and materials were not separated in meaning. But, the telegraph changed all that; a network of 'lightning lines' soon crossed the nation. The electrical messages that crackled along these wires were many times faster than the fastest trains, whose rails the telegraph wires paralleled" (p. 29).

The first significant impact of the telegraph was on the print medium. In 1844, newspapers seldom carried daily national and international news items. However, as Czitrom (1982) has noted, the telegraph "made possible, indeed demands, systematic cooperative news gathering by the nation's press" (p. 16). Therefore, the Associated Press (AP) was formed in 1849.

Once the principle was established, electronic modes of communication proliferated, and, compared to prior communication developments, the proliferation was amazingly rapid and pervasive. The first successful motion picture camera was developed in 1889, and *The Great Train Robbery* was released in 1903. In 1897, the first wireless telegraph was patented, and in 1906 the first successful radio system was demonstrated. In 1877, Thomas Edison filed the first patent for a "talk machine," and in 1925 the first electrical system for recording and playing back sound was perfected. In 1922, the first electronic television system was designed, and the first commercial television was broadcast in 1939. And, while the principles of contemporary computing theory can be traced back to the mid-1660s, the basic principles for electronic computerization were developed by Charles Babbage in the early 1800s, and the first electronic computer unveiled in 1946 by the Moore School of Engineering at the University of Pennsylvania (Chesebro & Bonsall, 1989, pp. 25–26).

The societal and individual consequences of these new electronic media systems are still emerging. Indeed, the shift from telecommunication (i.e., one-way transmission of messages from source to receivers) to interactive (i.e., two-way transmissions between source and receivers) electronic communication systems is less than 50 years old, yet, compared to print, the electronic media have already altered what is perceived, how it is perceived, and the kinds of understandings that these electronically generated perceptions can mean. Television and film function as apt examples. These media highlight how the electronic technologies have altered what is perceived as information. They also illustrate how theories of communication have had to shift from a focus on arrangement and style, the dominant rhetorical emphases of the print medium, to the new emphasis on invention in the world of electronic media. Chesebro (1989) has previously explained that

> the visual component of electronic media such as television and film highlights motion. The apparent motion within a frame, the movement embedded within the progression of shots, and the series and sequences of shots in one location and from one location to another location defines what is known (see, e.g., Harrington, 1973). The traditional notions of arrangement and style which characterize literate cultures are dramatically altered in such electronic media. The constant motion characterizing all electronic media reflects, not only metaphorically but literally, a search and quest, or what has been identified in classical rhetoric as a concern for invention. (p. 12)

The consequences of these shifts for understanding and knowing have been dramatic:

> While the knower and what is known are reunited in electronic media such as television and film, knowledge is separated from the lived experience. Television and film conceptions of "what is" report only what can be seen and heard; but more importantly, this visual and auditory conception of "what exists" is typically understood within a totally unrelated context (the home or the movie theater), a context which did not characterize the original situation. Context-defining influences are thus lost in the electronic culture. (Chesebro, 1989, p. 12)

The details of these new implications are explored in Chapter 6.

PRINCIPLES FOR CONSTRUCTING A COMMUNICATION HISTORY

The conception of communication history offered here is intentionally selective, and therefore, as we noted and anticipated at the outset of this chapter, this historical sketch admittedly provides both an incomplete and a unique view of communication history. We have deliberately *not* emphasized major political and military events, leaders, speeches, and strategies that have exerted influences upon why and how human communication has evolved as it has. We believe our approach is a useful way of identifying significant moments in the history of human communication. As our descriptions of this history suggest, our approach focuses upon how two basic dimensions of human communication, *culture* and *technology*, provide explanations for the evolution of human communication that other approaches are likely to slight.

By way of explaining this cultural and technological orientation, the assumptions of this approach will now be appropriately highlighted. Some of these assumptions are definitional, others methodological, while some posit theoretical conceptions of communication itself. To identify the nature of these assumptions as explicitly as possible, eleven specific claims are highlighted and numbered sequentially in the following pages. For clarity, we have found it convenient to deal initially with our assumptions about the nature of media systems, then to turn to how cultural systems are understood in this volume, and finally to consider how technologies and cultural systems interact.

1. *A medium of communication is an active determinant of meaning.* *Webster's* (1986) has defined a *medium* as "something lying in a middle or

intermediate position," "a middle way," as "something through or by which something is accomplished, conveyed, or carried on," and as "the material or technical means for artistic expression" (p. 1403). In traditional conceptions of the communication process, the medium of communication is frequently treated as a *channel*. In his classical conception of the communication process, Berlo (1960) proposed that the channel was one of the six "elements that all communication situations have in common," and he defined a channel as "a medium, a carrier of messages" (p. 23). Yet Berlo also conceived of the channel as an active determinant of the meanings humans receive when they communicate. He held that "the *choice* of channels often is an important factor in the effectiveness of communication" and that the channel is one of the "ingredients of communication" (p. 28). Specifically, he suggested that the channel of communication functioned as the "encoding and decoding apparatus" that allows "external physical" stimuli to be "translated internally," ultimately functioning as the "encoding" and "decoding" system for all human communication (pp. 63–64). Cast as the perceptual or encoding/decoding system that gives order to external stimuli, the channel or medium of communication cannot reasonably be treated as only or as merely a "conduit of a message." In terms of microanalyses, media systems such as television news formats can shape public attitudes and ultimately social policy (see, e.g., Altheide, 1991). In terms of macroanalyses, media systems can be conceived as "constructs" or "organized schemas or patterns of expectation within which events are construed or interpreted. Any event can make sense only in so far as it is ordered within the construct system" (Delia & O'Keefe, 1979, p. 161). In all, reasoning that "there is nothing innate in the human nervous system which gives us direct information concerning space," and that "there is no specialized space receptor," Strauss and Kephart (cited in Wachtel, 1978, pp. 376–377) have maintained that "every medium has a bias toward space and time. Each, in its own way, imposes an order and a coherence on the world. This is not only true of language, but of all the new technologies and techniques that present experience in symbolic form" (see also Carpenter & McLuhan, 1960).

2. *The medium of a message generates a different kind of knowledge than the content of a message.* Focusing on both the development and cultivation of learning, Salomon (1979) has maintained, "When a medium's messages are encountered, knowledge of two different kinds are acquired: information about the *represented world* and information about the *mental ability* used in gaining it" (p. 55). He has concluded, "Thus, while the *contents* of messages and experiences address themselves to one's knowledge and map upon one's knowledge base, the ways they are structured and presented address themselves to one's mental skills or abilities" (p. 55).

In this sense, the medium employed to present information deter-

mines how experiences are utilized and incorporated into the human being as mental skills or abilities. Salomon (1979) has likewise suggested that media affect not only the development, but also the subsequent cultivation of mental skills. He asserts, "Assuming that the major media of communication differ—to a smaller or larger extent—in their modes of gathering, selecting, and packaging, and presenting information, it becomes important to examine the psychological consequences of these differences" (p. 55). Salomon has additionally noted that "because there is reason to assume that differences in modes of presentation are associated mainly with differential employment of mental skills, their effects on the *cultivation* of such skills can become a focal point in research" (p. 55).

3. *Every medium of communication creates and presents a unique view of reality.* More detailed and complete surveys of neurophysiological research examining the relationships between media and *cognition* (see Glossary) are available elsewhere (see, e.g., Chesebro, 1984, pp. 117–118), but Krugman's (1965, 1971) neurophysiological findings function as an excellent example of the ways in which each medium creates and presents a unique view of reality. Krugman (1971) has argued that television is a "popular, interesting, and time consuming" (p. 3) but "low involvement medium," perhaps "five times" less involving than print, and a form of "passive" reception in which "huge quantities of information" are "effortlessly transmitted into storage" (p. 8). Labeling television a form of "passive learning" (p. 9), Krugman notes that television viewing generates fewer spontaneous thoughts, fewer links to the content of the viewer's personal life (p. 3), and "unformed and shapeless" responses (p. 8). Accordingly, television is linked to the formation of general "orientation and awareness" rather than to the particular role-specific behaviors, computational skills, and logical analyses created by other media (Springer & Deutsch, 1981, p. 14; see also Phelps, Mazziotta, & Huang, 1982, p. 142).

4. *A communication technology perspective emphasizes the format of the medium more strongly than the content of messages.* The perspective outlined here quite intentionally seeks to draw attention to the communication technologies themselves as powerful variables affecting human communication, and this approach is ultimately designed to suggest how critics can examine technologies themselves as conveniently and easily as they have traditionally examined the content of communication. Although content is not dismissed, attention is consciously shifted to the examination of the knowledge generated by technologies themselves, for technologies, long neglected in critical analyses, need to be examined as thoroughly as content whenever media systems affect human action. In this context, a rationale for the shift from a content to a media emphasis is appropriately articulated.

This rationale begins with the recognition that a significant shift in the nature of communication has been reported for several decades. Increasingly, it has been noted that a shift from a content orientation—with its emphasis on the ideational or substantive dimension of discourse—to a concern for form or medium—with an emphasis on image, strategy, and patterns of discourse—has been identified as a central feature of the information age.

Concomitant with the growing importance of media, the content of a message, in the eyes of many media critics, has lost its place as the most significant variable in creating and maintaining a communicative relationship. The declining importance of content has been discussed in terms of the corresponding increase in "pseudo-events" and "image politics" (see, e.g., Boorstin, 1961; Boulding, 1956/1961; Lippmann, 1925, pp. 13–14, 78–80, 178–186; Swanson, 1972).

Likewise, in communication and rhetorical theory, a series of methodologies have emerged that emphasize the form, rather than the content, of human communicative interactions. Unifying this tendency has been a growing recognition that human symbol using is socially constructed and that symbols themselves are conventional and arbitrary constructions that have no necessary relationship to environmental conditions (see, e.g., Berger & Luckmann, 1967; Hastings, 1970; Krippendorff, 1984; Scott, 1967, 1976). Blurring the distinction between "what is real" and "what is believed" (see, e.g., Wander, 1983), human communication has increasingly been treated as fictions, in the form of fantasies and narratives. Indeed, for the last two decades, the fictive has been the dominant object of study of rhetorical theory and criticism (see, e.g., Bormann, 1972; Fisher, 1984).

In the case of television, content has become, at best, ephemeral. It is questionable if viewers find any particular episode of a prime-time television series meaningful, and, based upon ratings, almost 50 percent of prime-time series are annually replaced (Chesebro, 1986a, p. 512). In fact, Jacoby and Hoyer (1980, 1982a, 1982b) have reported that the average American forgets or misunderstands 23–36 percent of televised information when asked to recognize information immediately after a viewing. Insofar as prime-time television series exert an influence on viewers, the effects are apparently due to the format of these series rather than any particular plot, conflict, or content (Chesebro, 1991). Misinformation analyses of readers of print communications have also suggested that approximately one-third of average Americans forget or misunderstand the content they read (Jacoby & Hoyer, 1987). Additionally, computer processing systems are virtually indifferent to the content they classify, sort, and calculate (D. Sanders, 1985, p. 11). Reacting to these "information machines," Bagdikian (1971) has concluded that "electronics suddenly short-circuited the ancient linkage of literacy and abstract intellectuality" (p. 10).

5. *Habitual use of specific media systems privileges certain worldviews, perspectives, orientations, or viewpoints.* The persistent use of any one medium as both a source of information and as a leisure activity reinforces a particular worldview or orientation. Reinforcing Wachtel's (1978) observations that each medium has a unique "bias toward space and time" and therefore "imposes an order and a coherence on the world" (p. 376), Carpenter (1960) has also reported that "the new mass media—film, radio, TV—are new languages," that "each codifies reality differently," and that "each conceals a unique metaphysics" (p. 162). He has maintained that "oral languages" create a worldview "composed of great, tight conglomerates, like twisted knots, within which images were juxtaposed, inseparably fused," in contrast to writing, which "encouraged an analytical mode of thinking with emphasis upon linearity" (p. 162). While admitting that he has found the "new media" to comprise nothing more than a series of "interruptions," nonetheless, offering a more descriptive analysis of these new media, Carpenter has noted that they offer a view of reality in which "short, unrelated" events can be placed in proximity to each other and thereby "become associated" with each other, "without antecedents or results" and therefore, in a strict sense, "never present an ordered, sequential, rational argument" (pp. 164–165).

6. *The production variables of a technology provide the most immediate and direct way of defining a communication medium.* McLuhan (1964) argued that the media were extensions of the human senses, and therefore are understood as a direct reflection of the goals and objectives of the human beings using them. However, Miller (1978) has aptly noted that media can function as either tools or technologies. Tools, he has noted, "extend the immediate biological capabilities" of human beings (p. 229); when media systems function as tools, they are appropriately understood as extensions of the self. However, Miller has argued, media systems can also function as technologies, losing their original purpose and beginning to function independently of the human being. We are particularly concerned about media systems when they function as technologies since by definition, it is no longer possible to characterize them solely in terms of the purpose attributed to them by human beings. Under such circumstances, alternative ways of defining the nature and functions of media technologies must be found.

Media technologies can be defined in any number of ways, but a production perspective focuses on the essential manufacturing activities that must be undertaken to generate physically a message uniquely defined by the technology. For example, to generate a televised message, at the very minimum, two dimensions—the *visual* and the *auditory*—must be present. The visual dimension must include the basic visual units, such as the shot (i.e., a single uninterrupted recording of a camera), a scene (i.e., a series of shots in one location and in the same apparent time period), and a sequence (i.e., a series of scenes unified by one location). To generate these basic visual units, a series of related decisions might also be made, including a discussion of the composition of the visual unit, image size, perspective, camera speed, lighting, continuity, and so forth. The auditory dimension must include the basic sound units, such as amplitude (i.e., loudness level), frequency (i.e., pitch), attenuation (i.e., duration), quality, harmonics (i.e., overtones), and interference. To generate these basic auditory units, a series of related decisions might also be made, including a discussion of the dialogue, local and background music, ambient sounds, dubbing, and so forth.

The production elements and processes constituting communication technologies affect what is perceived and apprehended. The production components of a communication medium are selective channels that convey only certain information about events, not all that is important about those events (Berg, 1972). As a result, only certain human sensory systems are stimulated, not all. Additionally, the production channels of a medium function as the most immediate context of message content, thereby generating a kind of information as important to apprehension as the information generated by content. In these senses, the production elements and processes define a communication technology, determine what people will and will not perceive and apprehend, constitute the basic "grammar" of a communication technology, and ultimately establish the structure that gives meaning to a message.

7. *Media systems can reflect and create their own culture*. The structural and production restraints of every medium of communication inherently select, emphasize, or encode certain external stimuli, but not all of the available stimuli. In so doing, the medium conveys one understanding, rather than other understandings, of what the available stimuli are. The singular understanding that can be conveyed by the medium ultimately functions as the foundation for valuing or judging the external stimuli. As Gumpert and Cathcart (1985) have reported, "people develop particular media consciousness because media have different framing conventions and time orientations" (p. 23). Accordingly, one would expect that as media systems change, orientations will change. Further, Gumpert and Cathcart have argued that

> persons are influenced by the conventions and orientations peculiar to the media process first acquired and relate more readily to others with a similar media set. Fifty and sixty year olds, for example, who have learned to process reality in terms of a logically ordered, continuous

and linear world produced by a primary print orientation feel linked in rejecting the world view of those whose electronic orientation is to a visual/auditory, discontinuous reality. On the other hand, eighteen to twenty year olds might feel removed from twelve to fourteen year olds because they cannot fully grasp the digitally oriented computer world. (p. 23)

8. *A cultural system may be examined as a discrete, unitary, and structural entity.* In its most commonly recognized form, a culture is a composite of societally shared standards; it is an integration of socially shared values that reflect the structures of a society. As conceived by Goodenough (1971), the study of culture directly involves an examination of those social structures that reflect, reveal, and emphasize the standards common to members of a societal system. The content of a cultural system is thus composed of the ways in which people have organized their experiences of the world, the system of cause-and-effect relationships, the hierarchies of preferences, and the recurring purposes used for achieving desired futures. As Goodenough concluded, "*Culture, then, consists of standards for deciding what is, standards for deciding what can be, standards for deciding how one feels about it, standards for deciding what to do about it, and standards for deciding how to go about doing it*" (p. 22). Thus, culture becomes the study of a set or pattern of societally shared standards.

9. *A cultural system is an active, not passive, entity. The standards constituting a culture are functional because they are used to make decisions, to determine which resources are amassed to realize decisions, and to choose what actions are and are not to be taken.* As Chesebro (1984) has previously explained,

The standards of a community function as both an evaluative and a control system. This system of values manifests itself as the lifestyles, customs, norms, institutions, ideologies, and principles which regulate forms of individual and social behavior and ultimately all modes of socially shared understandings. A cultural system is, by definition, an integrative system which unifies or at least establishes the standards for unification. In this sense, a cultural entity functions as an active symbolic and socialization system creating, reinforcing, and altering individual, sociological, and enculturation processes. (p. 114)

10. *Cultural systems can determine the meanings assigned to the content of messages.* The influence of culture as a determinant of the meanings assigned to an object, person, or event has already been established. For example, Liebes (1988) had five different cultural groups independently watch the same episode of "Dallas," and summarize the plot of the episode. Liebes reported that "the viewer retellings" fit into "three

different narrative structures" that she characterized as featuring "action" (i.e., "linear" or "sociological), "characters" (i.e., "segmented" or "psychological"), or "themes" (i.e., "paradigmatic" or "ideological"). Liebes concluded that "it cannot be taken for granted that everybody understands the programs in the same way or even that they are understood at all" (p. 289).

11. *Technologies and cultures interact and mutually define and determine human meanings.* A collage of variables affects the meanings attributed to and associated with an experience, person, object, or event. In some cases, as Liebes's (1988) study has demonstrated, the standards embedded within a cultural system apparently regulate how audiences react and understand. In other cases, as much of the evidence in this chapter indicates, the constraints built into a medium determine how audiences react and understand. Using her reading of an episode of the television series "Cagney & Lacey," Condit (1989) has argued that audience reactions may not be directly dependent on either the technology or their cultural system. She has suggested that audience reactions can be "constrained by a variety of factors in any given rhetorical situation. These factors include audience members' access to oppositional codes, the ratio between work required and pleasure produced in decoding a text, the repertoire of available texts, and the historical occasion, especially with regard to the text's positioning of the pleasures of dominant and marginal audiences" (pp. 103–104).

Any single analysis of audience reactions will inherently be incomplete. A communicative experience is far richer and more multidimensional than any single analysis of that experience can reflect and reveal. When Berlo (1960) characterized communication as a process, he noted that communication was "dynamic, ongoing, ever-changing, continuous. When we label something as a process, we also mean that it does not have *a* beginning, *an* end, a fixed sequence of events. It is not static, at rest. It is moving. The ingredients within a process interact; each affects all of the others" (p. 24).

Yet an analysis of a communicative experience arbitrarily isolates a communicative experience, assigning it a beginning and an end, and assumes that the communicative experience was understood by an audience within a single context and time frame. Accordingly, an analysis cannot capture all of the factors influencing audience reactions to and understandings of a communicative experience. Every analysis of audience reactions simplifies, ignores certain factors, and at best offers one perspective on audience reactions in the hope that the perspective helps to reveal the complexity of the human being.

Within this context, we claim that technologies and cultures interact and mutually define and determine human meanings. We do not *mean* to imply that technologies and cultures are the *only* variables determining the meanings humans attribute to their experiences. It is conceivable that particular communicative acts might be insightfully examined without any reference to the media system and cultural context in which communicative acts occurred.

At the same time, we are tremendously concerned about the decision of media critics, whether conscious or not, to ignore technology in assessing human communication. As Americans increasingly use television as their primary source of news, we find it extremely difficult to ignore Berg's (1972) observation that mass media coverage tends to distort the event, creating a "defect-ridden quality of reality" (p. 256) that differs markedly from the reports of direct observers. In addition, Berg has noted that "the rhetoric generated in response to mass media-produced reality differed from that which resulted from the direct observation of events or from word of mouth knowledge" (p. 258). Similarly, as Americans increasingly use prime-time television as their primary source of entertainment, we find it extremely difficult to ignore the demographic conclusion reached by several researchers that "the typical central television character does not reflect the American culture as it is" (see, e.g., Chesebro, 1979, p. 522, as well as his survey of literature on this issue). More profoundly, we are particularly concerned that communication technologies themselves, regardless of their content, are being bypassed as objects of study. Many of the new communication technologies, such as the computer, are intentionally designed to alter and change human inputs and generate outcomes that have little relationship to the original human inputs (see, e.g., Williams, 1982, p. 108).

Similarly, we are tremendously concerned about the decision of media critics, whether conscious or not, to ignore cultural systems in assessing human communication. We recognize that media criticism will always reflect the perspective of the critic generating the criticism (see, e.g., Black, 1965). At the same time, we are equally impressed with the conclusions we derive from Liebes (1988) and Condit (1989) that a critic's personal view and evaluation of a communicative act will not reflect, especially in a multicultural system, the orientations and understandings of others. No matter how well intentioned, there is a sense in which the descriptions, interpretations, and evaluations offered by a media critic are less useful, if not misleading, if diverse cultural contexts are ignored.

Decisions to emphasize the interaction between technology and culture have generated unique and revealing insights of human communication. Havelock (1963) has examined the consequences of the ancient Greeks' shift from orality to literacy, offering a profound understanding of the evolution of Greek thought. Moreover, while cultural systems differ, the results generated by other researchers suggest that Havelock's observations are apparently both reliable and valid. Havelock's conclusions are confirmed, for example, in Clanchy's (1979) examination of the shift from orality to literacy in medieval England and in Luria's (1976) analysis of the shift from orality to literacy in the Soviet Union in the late 1920s.

In all of these senses, we believe it is useful, if not essential, to examine how technologies and cultures interact and mutually define and determine human meanings. This belief is but one perspective, but we hope that this perspective is sufficiently insightful to influence the subsequent choices of media critics.

| CONCLUSION

In this chapter, we have suggested how human beings have evolved into the kinds of communicators they are. This overview has provided, we hope, a useful view of significant moments in the history of human communication as it revealed the perspective on and understanding of criticism that dominates this book. In this chapter, we identified prior approaches to understanding the history of human communication as the context for an approach emphasizing the role of technology and culture in the evolution of human communication. From this perspective, a brief history of human communication was provided. We concluded with an explicit discussion of the principles underlying this history, which are essential features of the paradigm of criticism presented in this volume.

CHAPTER 2

A World of
Communication
Technologies and
the Human Response

| THE COMMUNICATION REVOLUTION

Increasingly, we live, work, and play in environments created, sustained, and altered by and through communication. Our daily existence—from the moment we get up, as we work, during our leisure hours, and until we retire—is an unending series of messages, information bits, symbols, signs, warnings, commands, images, strategies, and artifacts. Among several other functions, we use these streams and cross-currents of meaning to define who we are, who others are, and what our environment is. In this sense, a communication revolution has dramatically and decisively controlled the United States since World War II.

We have also encountered a communication revolution in another sense. Certainly, each of us continually confronts systems and networks of verbal and nonverbal messages in our face-to-face interactions. However, the nature of our communicative interactions is also changing in revolutionary ways. Instead of interacting with people face to face, we now increasingly communicate through artificial channels, technologies, tools, mechanisms, and machinery. Accordingly, we receive an overwhelming number of messages from televisions, videocassette recorders (VCRs), radios, newspapers, magazines, compact discs (CDs), records, business notices and reports, mail, telephones, government forms, scientific reports, advertisements, watches and clocks, menus and recipes, billboards, books, and various kinds of computer systems.

The issue is no longer whether or not a communication revolution *exists* but what this communication revolution *means* to individuals and various sociocultural groups as well as to the diverse institutions and structures in the United States.

In this chapter, we examine the different meanings that communication technologies have exerted on people. By contrast, in the first chapter, we traced changes in communication technologies over time. Here, however, we particularly focus on the different ways in which these technological changes have been understood, studied, or constituted as socially meaningful experiences. Whenever one describes, interprets, and evaluates communication technologies, several critical issues emerge: (1) how pervasive the systems are in society; (2) how people are using these systems; (3) how cultural systems are adjusting to these media uses; and (4) what the values of these technologies are to the individual, cultural groups, and society.

In the hope of understanding the nature, functions, and values of communication technologies for the individual and for society, we examine four approaches to communication technologies. First we consider what we would know about communication technologies if we measured the frequency with which media systems are used, an approach we have identified as the *media consumption approach*. Next, we examine why people use media systems and what kinds of gratifications they receive from such usage, a *uses-gratification approach*. Following this approach, we explore efforts designed to describe how cultural systems are affected by media systems, a *cultivation approach*. We conclude this chapter by examining how communication technologies affect the value systems of individuals, cultural groups, and society, a *critical approach*. We illustrate this critical impulse toward media by focusing on the analyses provided by Marx (1964), Sanders (1991), and Perkinson (1995).

Overall, then, the meaning of this communication revolution has been a subject of concern for a host of communication researchers and communication critics. We want to offer a sampling of some of these communication research and critical approaches. This minisurvey provides a convenient introduction of several of the basic facts, issues, and questions that emerge regarding the meaning of the communication revolution. We begin with a fundamental but essential question: *What are the most frequently used media?*

THE MEDIA CONSUMPTION APPROACH

Perhaps the most common way of explaining the communication revolution is by counting the *numbers* or providing *frequency* measures of communicative exchanges and contacts with communication technologies. Providing only an introductory description of communication technologies in human society, this approach predominantly treats human beings as media consumers and societal institutions as media systems. Although the approach is particularly useful for noting what people do, it slights any explanation of why people are communicating the way they are. Nonetheless, as a point of departure, this approach can be useful and offer a basic view of the human condition.

One of the most comprehensive examinations of media usage has been provided by DeFleur and Ball-Rokeach over a 30-year period in a series of five editions of *Theories of Mass Communication*. In this volume, the authors have provided a series of frequency tables that plot human use of communication technologies over time. Although these tables do not explain why certain changes have occurred in media usage, a wealth of preliminary information is contained in them. For example, the following "facts" about the communication revolution emerge:

- *Daily newspapers:* Daily newspaper circulation per household reached its peak at 1.05 newspapers per home in the year 1965 and has consistently declined since that time (1989, p. 59).
- *Photography:* The number of photographers per 100,000 people in the United States reached its peak in the year 1900 and has steadily declined since that time (p. 73).
- *Film:* Weekly film attendance per U.S. household reached its peak in 1936 and has declined since that time (p. 81).
- *Radio:* The average number of radio sets per U.S. household reached its peak of one radio per home in 1936 and thereafter continued to grow until each American home had some 5.5 radio sets in 1985 (p. 107).
- *Television:* Between 1950 and 1955, the television became a national communication technology in the United States. In 1950, 9 percent of U.S. households had a television. In 1955, the number of televisions per U.S. households had grown to 78 percent. Now, virtually all American households, some 98 percent, have a television set (p. 114).

In an equally important way, virtually every measure we have indicates that the number of messages is increasing exponentially. Paradoxically, it becomes extremely difficult to convey, in a comprehensive manner, how rapidly this *rate* of change has occurred. If you have ever wondered why American society seems preoccupied by the technical, consider how dramatically the number of technical reports produced in the United States has increased during the 20-year period from 1968 through 1987: that number doubled every year from 1968 to 1977, quadrupled from 1978 to 1982, and increased by a factor of eight from 1983 through 1987 (Chesebro & Bonsall, 1989, p. 14).

In this regard, there is some evidence that this communication revolution has even affected our personal lives. Chesebro and Bonsall (1989) calculated that "the combined hours devoted just to watching television, reading, and talking on the telephone constitute 39 percent of all of the available hours within an entire year. If the time devoted to sleep is discounted, we spend 58 percent of our waking hours engaged in these media activities" (p. 15). In all, each of us now processes more information from more sources more efficiently than in any previous era.

Beyond dramatically affecting our personal life, the communication revolution now governs the primary economic and power dimensions in U.S. society. Almost 15 years ago, Americans were shocked by Fowler's (1983) report that, for the first time in its economic history, more than 50 percent of the U.S. gross national product was attributable to the "development of data, exchange of information, manipulation of ideas and the transfer of numbers" (p. D20). Yet, in a larger context, as Table 1 indicates, since 1860 an increasing portion of Americans have been, and will continue to be, employed in the information sector of the American labor force.

Correspondingly, as the economy increasingly relies on the exchange of information as the definition and foundation of income, the foundation of power has increasingly shifted from force and money to information. In terms of politics, Toffler (1990, p. 3) has argued that "we live at a moment when the entire structure of power that held the world together is now disintegrating" when "a radically different structure of power is taking form" in which power is shifting from its traditional base of "violence" and "wealth" to a system based on the "knowledge" generated by the "new technologies".

Beyond its economic and political significance, this communication revolution also has had a profound impact on each individual's life. In one sense, we have had to "remake" ourselves into information processors. Although useful, if not essential, in multiple contexts, the constant stream of information cannot be absorbed. There is simply too much information for any given individual to deal with in any useful way. Wurman (1989) coined the expression *information anxiety* to capture one reaction to the ever-increasing information overload each of us experiences. Selective attention and selective perception must be applied consistently if information overload is to be prevented, and if individuals are to have the sense that they personally—rather than streams of data—control their own lives.

TABLE 1. Four-Sector Aggregation of the U.S. Workforce by Percentage, 1860–2000

Year	Information sector	Agriculture sector	Industry sector	Service sector
1860	5.8	40.6	37.0	16.6
1870	4.8	47.0	32.0	16.2
1880	6.5	43.7	25.2	24.6
1890	12.4	37.2	22.3	22.3
1900	12.8	35.3	26.8	25.1
1910	14.9	31.1	36.3	17.7
1920	17.7	32.5	32.0	17.8
1930	24.5	20.4	35.3	19.8
1940	24.9	15.4	37.2	22.5
1950	30.8	11.9	38.3	19.0
1960	42.0	6.0	34.8	17.2
1970	46.4	3.1	28.6	21.9
1980	46.6	2.1	22.5	28.8
1990	43.0	1.4	21.3	34.3
2000	42.5	1.3	20.9	35.3

Note. Data for the years 1860 through 1980 are derived from Porat, with Rogers Rubin (1977). Projections for the years 1990 and 2000 are derived from the definitions and measurements techniques contained in Porat (1977), and specific data contained in Hawken (1983) and Rogers Rubin, Taylor Huber, and Lloyd Taylor (1986). For a discussion of factors that have recently constrained the development of the information sector, see Dizard (1989).

In all, this "media consumption" approach provides the basis for concluding that we have created and live in an information society and age. Functioning within this context and drawing conclusions that reflect this orientation, Dizard (1989) has also underscored the corresponding importance of criticism in such an era:

> Of all of the changes taking place in our time, none has more profound effects than the new ways in which we communicate with one another. For the first time in human history, there is a realistic prospect of communications networks that will link everyone on earth. . . .
> There are no historical precedents to guide us through this momentous change in how we deal with one another. No one knows the ultimate effects of this new order of instantaneous links. Most predictions tend to stress the advantages of what is usually described as "bringing people closer together." This is undoubtedly a positive gain,

although there is a dark side to the picture, given the destabilizing factors brought about by the fragmentation of old patterns of personal and tribal isolation. (p. 1)

As Dizard implied, if we are to survive in the information society, there is a profound sense that we will survive only if we develop skills as critics. Every living organism must respond to its environment to determine what is healthy and supportive, and what is destructive. The same is also true in an information society. Every living organism must be a critic of information, able to describe, interpret, and evaluate its information environment, even when that environment is a made up of streams of contradictory data. As Burke (1935/1965) observed, "all living organisms interpret many of the signs about them . . . [t]he very power of criticism has enabled humans to build up cultural structures so complex that still greater powers of criticism are needed" (p. 5).

THE USES-GRATIFICATION APPROACH

The uses-gratification approach seeks to describe how and why people use media systems as they do. Blumer (1980, p. 202) has aptly observed that "the uses and gratifications approach came most prominently to the fore in the late 1950s and early 1960s at a time of widespread disappointment with the fruits of attempts to measure the short-term effects on people of their exposure to mass media campaigns" (p. 202). Uses-gratification researchers particularly assume that audience members actively and individually use and/or are gratified by media systems in different ways. Accordingly, uses-gratification researchers seek to understand media effects by identifying the ways in which individuals use the media and/or are gratified by media exposure. The most common uses-gratification research strategy assumes that if you want to know how people are affected by media systems, ask them. Becker (1980) has aptly noted that the most common strategy for conducting uses-gratification research "is to rely on reports from the audience members" (p. 229).

To date, uses-gratification research has suggested that people use media for one or more of four basic reasons: (1) *escapism*—to avoid ongoing reality systems; (2) *reality exploration*—to secure basic information and to understand the world in which they exist; (3) *character reference*—to find suitable models for their own lives; and (4) *incidental learning*—a kind of miscellaneous category in which it is recognized that each individual may use or be gratified by media for very different, personal, and unique reasons.

Uses-gratification research has generated some extremely disturbing issues. For one thing, there appears to be no correlation between the content and form of media systems and how people respond to them. In their study of "miscomprehension of televised communication," Jacoby, Hoyer, and Sheluga (1980) reported that when some 2,700 respondents were asked to view two television communications and answer six quiz items "about the main idea of the commercial or film clip" (p. 52) they had just seen, 29.6 percent of the items were not correctly answered. They concluded that a "large proportion of the American television viewing audience tends to miscomprehend communications broadcast over commercial television" (p. 89). They specifically reported that "the vast majority (96.5%) of the 2700 respondents in this investigation miscomprehend at least some portion of the 60 seconds worth of televised communication which they viewed . . . [t]he average amount of miscomprehension associated with each of the 60 test communications was 29.6%" (p. 52). Overall, Jacoby, Hoyer, and Sheluga concluded that "approximately 30% of the relevant information content contained within each communication was miscomprehended" (p. 52). Additionally, it should be noted, Jacoby and Hoyer (1987) have also drawn similar conclusions regarding print communication.

For the uses-gratification researcher, these miscomprehension findings are particularly disturbing because the research design assumes that the uses and gratifications of media consumers are somehow related to what the researcher understands the content of the media system to be. Miscomprehension studies deny that there is any necessary relationship between what media consumers derive from media systems and the "main idea" of a media system itself. Accordingly, Windahl (1981) has reported that "the criticism most commonly expressed" regarding uses-gratification research is that the approach is "too individualistic in method and conception which makes it difficult to tie to larger structures, . . . relies to a high degree on subjective reports of mental states and is therefore regarded as too 'mentalistic,' . . . assumes that media behavior is based on conscious or rational choice, which goes against research results saying that media use is habitual and nonselective, . . . [and] shows little or no sensitivity to the substance and nuances of the media content itself" (p. 175).

These judgments of uses-gratification research are compelling, but it should be noted that the approach itself continues to appeal to many communication researchers because the approach promises to address the question of *why* people are as deeply involved with communication and media systems as they are. Accordingly, some researchers have offered reconceptions of the uses-gratification approach. They have asked how to link media effects research and media uses-gratification research. In this regard, Rubin (1993) has reported that new and more "elaborate models of media effects" may be required that consider "social and psychological attributes, motivation, attitudes, behavior, and outcomes" (p. 103). To date, Rubin has concluded, "we have just touched the surface on understanding the role and outcomes of mediated communication for individuals and societies" (p. 103). Accordingly, we would encourage the exploration of more elaborate models linking media effects and uses. We find it extremely important to consider why people use media the ways they do. However, until a more compelling system is developed, we would recommend considering alternative schemes such as Gerbner's theory of media cultivation, to which we will now turn our attention.

| THE CULTIVATION APPROACH

The cultivation approach seeks to describe how culture systems have been affected by media systems. In essence, the effort of these cultural indicator researchers has been to determine if media systems affect cultural norms and understandings. For example, Gerbner and his colleagues (e.g., Gerbner & Gross, 1976; Gerbner, Gross, Eleey, Jackson-Beeck, Jeffries-Fox, & Signorielli, 1977; Gerbner, Gross, Morgan, & Signorielli, 1982) have argued that individuals who view television excessively believe that there is more crime than crime statistics demonstrate, while those who watch less television tend to report more accurately the extent of crime within their neighborhoods.

One conclusion that these studies would seem to warrant is that media systems may be a direct basis for the social construction of reality and the social issues (agenda setting) that people respond to and treat as real. Accordingly, a direct relationship is believed to exist between media use and the development and evolution of cultural systems.

Yet the approach makes certain assumptions that many people would find disturbing. First, it assumes that media systems exist independent of cultural systems. Cultivation theory assumes that media systems function as causal systems external to cultural systems. Media systems cannot exist independently of cultural systems. Media systems can only be a product of and reflection of the cultural, political, economic, and social systems that created and sustained them in the first place.

Additionally, cultivation theory appears to ignore the power of multiple causation in its conception of communication experiences. In this view, nonmedia systems can be as powerful as media systems in terms of their ability to shape cultural norms. Indeed, for many, media systems themselves are multicausal systems insofar as they function as intertextual systems that mutually and simultaneously influence each other. In this regard, Faber, Brown, and McLeod (1986, pp. 556–558) have been able to demonstrate that television is likely to affect only the secondary, not the primary, socialization process during adolescence. In other words, although adolescents will learn what society values, such as beauty and handsomeness, from television (secondary socialization), television is unlikely to affect dramatically sex-role determinations (primary socialization), simply because, as Faber, Brown, and McLeod have noted, "recent research on the concept of androgyny, or the blending of roles based on gender, suggests that sex roles are not a unidimensional construct" (p. 557). In this regard, given a complex and interrelated world, then, we may use television to characterize our mode of sexuality, but we are unlikely to select a basic mode of sexuality based on exposure to television. Sex-role determination is affected by far more variables than simply television viewing.

In a larger sense, cultural indicator research faces severe limitations because it overstates the power of media systems as independent variables and understates the influence of competing socialization variables. Thus, while hoping that new research findings may "help motivate researchers [to] take up the question of cultivation effects," Shrum (1995) has concluded that "this line of research has languished somewhat, it seems, in the last few years, perhaps because of the difficulty in addressing the very fundamental criticisms that have plagued it" (p. 422). Given these conditions, we find it potentially useful to examine how critics have examined media systems and communication technologies.

| THE CRITICAL APPROACH

Critical judgments of media systems can be offered in any number of ways. A statement of preference for one style or presentational mode rather than another (e.g., "I liked the book more than the movie") may provide the foundation for a critical view of a media system, or media criticism may include a far more sophisticated and complex description, interpretation, and evaluation of a communication system, such as when, for example, we seek to determine how and toward what ends computer systems affect the human condition.

Given the large number of ways in which media criticism might be defined, we can more effectively begin by considering an example of media criticism that few would deny functions as an excellent, if not classic, case of media criticism. In this regard, we survey critical analyses of communication technologies that reflect the richness and potentialities that a critical approach might provide.

We begin with Marx's (1964) analysis in *The Machine in the Garden*, which was published over 30 years ago, but which reflects and isolates some of the enduring issues that media critics have grappled with for several centuries. Marx argued that the American experience had been undergoing a profound transformation, one that involved more than machinery and social reorganization. The transformation also involved the American spirit, how Americans were talking about themselves and their environment, and ultimately how Americans were to understand themselves. In *The Machine in the Garden*, Marx explored the pastoral ideal that had characterized the American vision, but he did not simply argue that Americans were attempting to restore their pastoral ideal, *sans* machinery. Nor did he argue that technologies were replacing the pastoral ideal. Instead, he maintained that Americans were redefining their vision to incorporate machinery and technology into it. As Marx (1964) saw it, an effort was underway to create a "rhetoric of the technological sublime" (pp. 195–207, 214, 230–231, 294–295). In making his argument, Marx developed a host of conceptions that form an important foundation for this volume.

Obviously, Marx's analysis was an early example of media criticism. His thesis went beyond an effort to alert others to the existence of new and emerging technologies. The power of his analysis resided in his claim that new technologies were fostering a new way of talking about ourselves, others, and our environment. Marx identified his object of study as the "cultural symbol," which he defined as "an image that conveys a special meaning (thought and feeling) to a large number of those who share the culture" (p. 4). Cultural symbols were examined to reveal a conception of "the general cultural" and the "contradictions" that are a "way of ordering meaning and values that clarifies our situation today" (p. 4). New technologies were generating a new way of thinking, that Marx identified as a "rhetoric of the technological sublime." In this regard, he was acting as a critic of a system that functioned, in part, as a communication technology, because he saw the new technologies as message generating. Emphasizing the symbolic implications of the rhetoric of the sublime, Marx held that the introduction of technologies in the United States functioned as a purifying moment of awe in which new technologies became a way of realizing a kind of spiritual, moral, and intellectual perfection or nobility. At the same time, he held that this effort was unsuccessful, because the effort generated "a powerful metaphor of contradiction" (p. 4) in the American vocabulary and culture that was "helping to mask the real problems of an industrial civilization" (p. 7).

As Marx concludes his analysis, he suggests that perhaps the artist possesses the orientation necessary to link the pastoral and technological in ways that would console. Yet as we examine Marx's final statements, it becomes clear that he believed, in the end, that the symbolic contradictions between the pastoral and technological had not and would not be resolved by the artist. Marx assumed that the pastoral–technology dichotomy was a useful symbolic construct, and he continued to believe that the tension it generated could be resolved. The concluding sentence in *The Machine in the Garden* is most revealing: "The machine's sudden entrance into the garden presents a problem that ultimately belongs not to art but to politics" (p. 365).

Several of Marx's descriptions, interpretations, and assessments are apt illustrations of the nature of a critical approach to communication technologies. Indeed, because both of us were formally trained as rhetoricians, we are particularly impressed by Marx's emphasis on the "rhetoric of the technologically sublime." Like Marx, we are convinced that a powerful rhetorical impulse continues to dominate the American discussions of technologies. Marx has provided a historical construction of the American experience that suggests a culture of technology has been created in the United States and that Americans continue, almost desperately, to impose symbolic frameworks on technologies, seeking new ways of describing, interpreting, and evaluating them.

Yet we are less convinced that Marx's "sublime" is the most appropriate concept to characterize the American need to reconceive technologies. Americans have consistently cast the "garden" (or, what is "natural," "human," and "humane") in opposition to the quest for technological development. This dialectical tension has characterized and probably will continue to characterize the American dialogue about technology into the foreseeable future. For example, Aden (1994) maintained,

> As we move from the industrial age to the technological age, the intrusion of the machine into the garden becomes complete. In fact, mass media technology "rivals" God in its ability to spread messages and monitor our behavior. Additionally, production in a technological age is even less necessary than in an industrial age since services rather than goods become the chief commodity. . . .
>
> The response to this cultural transformation, however, is resistance rather than acceptance. Some cultural evidence suggests that Americans are seeking to move back into the garden, using it as a place of refuge and spiritual rejuvenation. . . . They marshall evidence indicating that more individuals are moving to wilderness areas, a trend reaffirmed by Californians' exodus to more rustic parts of the country. Moreover, Americans appear to be paying more attention to their spiritual needs, whether it be through church attendance, praying, or involvement in spiritual organizations. Finally, the percentage of the work force that is self-employed has grown steadily since the 1970s, the opposite of the trend during most of the 20th century. (p. 309)

Given the changes noted by Aden, the tension between humane and technological tendencies, increasingly defined by communication concepts such as "The Communication Revolution," constitutes a context with which any volume of communication technology criticism must deal.

At the same time, it is no accident that media critics such as Marx and Aden have been concerned with the role of machinery in the human environment. These critics have responded to the ways in which technologies of communication have affected American culture. In this regard, there are a host of critics whose preoccupation has been the technologies of human communication systems.

Sanders (1994) argued that the increase in "violence" in U.S. society is related directly to the decline in orality and "the silencing of the written word" because of increasing use of "electronic media" (p. 237). Many have found Sanders's claim to be controversial, especially because of his reliance on Freudian concepts and his sexist allusions. However, as a media critic, Sanders believes the relationships among oral, written, and electronic communication systems can and should be used to explain the human condition. Describing his own orientation, Sanders (1994) has maintained, "I have described in this book the historical and traditional way that people journey into literacy without leaving their early familiarity with orality behind. I have also tried to show that the traditional route has been destroyed, and the borders closed to future travel. I have outlined the detours that young people have taken into gangs and violence and drugs, in their attempts to claim a voice" (p. 237). In this context, Sanders has concluded, "Without proper immersion in orality, however, no one can really hope to set out on that journey, let alone reach a destination in literacy" (p. 237). In Sanders's view, literacy must begin at home. He has observed, "I use the passages from D. W. Winnicott as an epigraph to this chapter to emphasize that the process of literacy starts with the baby's mouth working the mother's breast. Illiterates show up in the classroom; dropouts show up on the streets. But they represent no more than the symptoms, the results, of a much deeper, more profound relational problem that begins in the home" (p. 237).

In this view, children have been affected from the earliest stages of their infancy. In Sanders's view, "Modern technological innovations like baby bottles and formulas make it possible for babies to reject their mothers' *alma ubera*. What might be called a thickening of the child's development through contact and deep nourishment ceases at that moment of the separation of the baby from its mother" (p. 240).

For Sanders, a profound revolution must occur in the United States if literacy is to be achieved. He has maintained,

To effect changes in literacy levels in this country, nothing short of the ideal can be tolerated, because the ideal describes what used to be called normal. Everyone who reads this has a choice: The eradication of drugs and violence and illiteracy is possible, but not through some governmental program, or some bureaucratic agency. Bureaucracies and agencies have caused the problem. The solution can come only if teachers and parents and administrators first hold a vision of what life should look like, and then be willing to work to realize it. (p. 240)

Ultimately, then, for Sanders, the relationship between the family and literacy must be reestablished. He has concluded,

Like all relationships, the one between family and literacy is reciprocal. Literacy has kept the family alive—through discussions, critical analysis, stories, arguments, and conversation. Insofar as the computer has helped to erase the inner core of the human being—conspiring, that is, in the obliteration (*ob-littera* = "the erase of letters") of stories and storytelling—it has hastened the destruction of the family. The family wraps itself around the dynamic core of orality. Each member of the family, like a member of a tribe, carries those shared stories with them. The family narrates a life together. (p. 241)

Offering an alternative view that challenges several of Sanders's assumptions, Perkinson (1995) has isolated positive sociocultural impacts of the technologies of communication. In *How Things Got Better: Speech, Writing, Printing, and Cultural Change*, he posited the currently unpopular opinion that Western culture has improved over time, and that the media or technologies of communication have played a pivotal role in helping to make things better. Perkinson has maintained that human speech, when it first emerged, enabled people both to understand better the world they inhabited and to construct political, economic, and social arrangements to improve their lives. With the invention of writing in Sumer, and especially after the invention of the phonetic alphabet in Greece, Perkinson has argued that people were able to devise even better understandings and improved social arrangements. The invention of the printing press in the late 15th century led, in Perkinson's view, to the creation of the modern nation-state, capitalism, the open society, and modern science.

Finally, Columbia University's Freedom Forum Media Studies Center recently released a special issue of its *Media Studies Journal* that focused solely upon "Media Critics" (1995) and vividly illustrated how diverse their political positions and social orientations can be. This special issue "identified nine critics whose judgments are noticed in the editorial offices of major newspapers and the studios of network television" (p. v). The editors emphasize that no single political or social perspective unifies media critics: "What the breadth of their skills suggests is that the media should be criticized in many ways and from many points of view. No single omniscient critic can be expected to provide comprehensive criticism— the job is just too large and too varied" (p. v). At the same time, all of these "critics with clout—nine who matter" share a common concern for the ways in which the technologies of communication are used as social forces (p. v).

Viewed this way, and although rarely recognized as such, the critical analysis of communication technologies has become a pervasive form of communication in U.S. society. The critics generating these analyses admittedly are diverse, yet they share a common interest in the media and in the technological systems human beings use to communicate. Thus, as we see it, media criticism is a necessary product of and response to a specific situation or context, the communication revolution.

In our view, the communication revolution is a product of two discrete trends that have converged.

The first trend has been the dramatic increase in the use of communication, not only in the primary economic and political institutions sustaining the American culture, but also as the primary vehicle by which individual selves are created and by which the individual as a social creature is sustained through communicative interactions with others. In other words, communication is now the means, perhaps the only means, by which individuals are able to interact and understand others as well as themselves.

A second trend is reflected in the American tendency to increasingly use technologies to facilitate all changes. As Marx (1964) demonstrated, technological developments have affected not only the U.S. modes of production, but also the American spirit. Mechanization and technologizing developments inspire Americans and reflect part of the quest that motivates and drives them.

In our view, the communication revolution is a product of these two discrete but interacting trends. Americans have fostered the importance of communication, viewed communication as critical to understanding who they and others are, and increasingly used technologies to enhance the understandings they derive from communication. Hence, for example, television news is the primary source of information used by Americans to know "what is happening," and the telephone is a regular feature of over 95 percent of U.S. households. Communication systems in the United States are mediated through technologies, and some of these technologies serve only communicative ends. Thus, they are now known as "communication technologies." When the trends toward increasing reliance on communication and increasing use of technology merged, the communication revolution came into being. The impact of this merger is only beginning to be felt.

| SELECTING AN APPROACH

We have surveyed four different approaches for examining communication technologies. For most of us, there is a tendency to "pick one way rather than another" when given a range of choices and options. Given that we each have a bias about how to do things, there also is a tendency to select the one procedure that seems most satisfying to us.

We cannot deny that we have made a choice among these approaches. We are, by inclination and practice, critics. We find the role an important role that allows us to examine and probe what is, classify and determine what a subject means to different groups of people, and render judgments about the value of a subject in ways that we hope help others, enhance the quality of life, and promote necessary changes in society where they are required.

Yet we would not advise that a commitment to any one of these approaches be made. For those new to the study of communication and of communication technologies in particular, all four of the approaches we have discussed will be useful. Whenever one describes, interprets, and evaluates communication technologies, one needs to know how pervasive the systems are in society (the media consumption approach), how people are using these systems (the uses-gratification approach), how cultural systems are adjusting to these media uses (the cultivation approach), and what the values of these technologies are to the individual, cultural groups, and society (the critical approach).

Accordingly, although choices may ultimately be made, we especially advise that those new to the study of communication delay such choices as long as possible. Ultimately, we believe, a commitment may be made to one of these approaches, but the contributions of all four approaches will continue to intrigue and shape understandings of communication technologies.

| CONCLUSION

Since World War II, a new social context has been emerging for every human being. This social context has been created by the types of communication we increasingly have been using. Both the number of communication systems and the quantity of information generated by each of them has been increasing exponentially, creating a communication revolution that now dominates the United States' economic, social, and political systems. As these new communication systems have emerged, corresponding interest in communication, media, and techno-logical criticism has occurred.

In the hope of understanding the nature, functions, and values of communication technologies for the individual and for society, we examined four approaches to communication technologies. First we considered what we would know about communication technologies if we measured media consumption, the frequency with which a media system is used. Next, we examined the uses-gratification approach to explain why people use media systems and the gratifications they receive from such usage. Following this approach, we explored the cultivation approach that describes how cultural systems are affected by media systems. We concluded the chapter by examining a critical approach that seeks to describe, interpret, and evaluate the human–technology relationship as a communication system. We illustrated the impulse toward media criticism with the analyses of Marx, Sanders, and Perkinson.

PART II The Critical Moment and the Critic's Method

W e can be attracted and repulsed by critics and what they do. Critics can attract us, because they frequently display honesty, conciseness, and courage. Reflecting these traits, in perhaps the most famous book review ever written, Francis Jeffrey said of Wordsworth's *Excursion:* "This will never do." On the other hand, we may distrust the abilities and talents of critics. James Russell Lowell once wrote,

> Nature fits all her children with something to do,
> He who would write and can't write, can surely review.

We will never dissolve all the mysteries and contradictions that surround critics. But we can systematically describe and therefore better understand the discourse they have generated.

Part II, "The Critical Moment and the Critic's Method," characterizes the choices, objectives, and discourses of critics. It is composed of Chapter 3, "The Critical Process."

Chapter 3 examines the nature of criticism in general. This chapter proceeds along three lines. First, some changes in what is viewed as the object of study for criticism are experienced. Second, the notion of "mediated communication" is specified and isolated in a way that more easily allows a critic to analyze communication technologies. Third, 11 features of criticism are identified. Fourth, to provide an indication of the more concrete efforts of critics, some of the specific purposes guiding previous media criticisms are summarized. Fifth and finally, *technocultural dramas* are isolated as a potential object of study for critics. These technocultural dramas are the product of the dominant communication technologies of a social system in dynamic interaction with that system's culture. Technocultural dramas influence how social systems operate as well as the lifestyles and societal organizations of successive generations. A model is proposed for systematically describing and identifying the dimensions of technocultural dramas.

CHAPTER 3   The Critical Process

C ritical analyses of speeches, printed communications, films, television programs, and computers regularly appear in newspapers and magazines, and on television and radio. Now pervasive, these analyses cross traditional media boundaries. A critical review of the film Batman Forever can appear in a financial newspaper such as *The Wall Street Journal* (e.g., Salamon, 1992). Morning television shows, such as "The Today Show," frequently review printed materials such as books and calendars. The politically elite *New York Times,* committed to printing "all the news that's fit to print," regularly publishes critical analyses of computer hardware and software every Tuesday and Sunday. In all of these ways, media criticism has become a ubiquitous dimension of American life.

Within the academic world, analyzing media also has become a significant and increasingly pervasive kind of scholarship. Some academic journals have specialized in media analyses since their inception. For example, the *Journal of Film and Video* began publication in 1947, the *Journal of Broadcasting and Electronic Media* in 1955, *Media, Culture and Society* in 1978, and *Critical Studies in Mass Communication* in 1984. Some of these publications, such as the applied telecommunications and information technology journal *Telematics and Informatics: An International Journal,* founded in 1984, specialize in highly selective areas of media systems. Nonetheless, at this point in time, virtually every communication journal includes analyses of media within its pages. Analyzing media has become ubiquitous in contemporary American communication studies.

In this chapter, we describe the choices and procedures a critic undertakes when describing, interpreting, and evaluating communication technologies. Specifically, we hope to create an appreciation and understanding of what critics examine, how they analyze media, the reasons for undertaking such analyses, and how critics deal with communication technologies as determinants of cultural systems.

This chapter is guided by five objectives. First, we isolate some of the changes involved in analyzing communication technologies. Second, we identify an appropriate object of study when analyzing communication technologies. Third, we isolate the principles a critic uses to analyze media. Fourth, we isolate some of the specific purposes and objectives that emerged when communication technologies were described, interpreted, and evaluated in previous studies. Fifth and finally, we describe the specific critical framework that can be used to isolate and characterize the technocultural dramas or cultural systems generated by communication technologies. (See Glossary under "Technologies as Communication Systems.") Thus, overall, in this chapter, we identify what we mean by *technocultural dramas* and how they can be described, interpreted, and evaluated. Chapters 4 through 6 provide extended examples of how types of technocultural dramas can be conceived and derived as well as characterized and evaluated. Accordingly, as we suggested at the outset of this chapter, our purpose here is especially pragmatic. Our effort is to reveal some of the criteria used and decisions made when principles of criticism are applied to the analysis of communication technologies. We begin this analysis by isolating some of the changes that have occurred in the study of mass communication; these new changes have altered what it is and how critics view and analyze communication technologies. The exploration ultimately allows us to isolate the appropriate object for a critic's analysis.

A TRANSFORMATION IN THE STUDY OF COMMUNICATION TECHNOLOGIES: IDENTIFYING THE PROPER OBJECT OF STUDY

Under the rubric of "media studies," critics might examine any number of objects. For some, mass communication systems are believed to be the proper object of study for those examining communication technologies. For others, mass communication can no longer define the domain of media critics.

The Shift from a Mass Communication Orientation

Traditionally, mass communication systems are believed to be equivalent to or an operational definition of communication technologies. Therefore, they are treated as the most appropriate object of study. However, we find such an equation limiting, and we believe that critics make a fundamental mistake if they isolate mass communication systems as the appropriate object of study. In defense of this position, a more precise definition of *mass communication* is an apt point of departure. In this regard, McQuail (1987) has identified the "main features of mass communication" in these words: "The source is not a single person but a formal organization, and the 'sender' is often a professional communicator. The message is not unique, variable and unpredictable, but often 'manufactured,' standardized, always multiplied in some way. It is also a product of work and a commodity with an exchange value as well as being a symbolic reference with a 'use' value" (p. 31). For McQuail, these features also possess specific causal interrelationships: "The relationship between sender and receiver is one-directional and rarely interactional, it is necessarily impersonal and often perhaps 'non-moral' and calculative, in the sense that the sender usually takes no moral responsibility for specific consequences on individuals and trades the message for money or attention" (pp. 31–32).

There is little question that mass communication, defined in this fashion, does exist, function, and exert influence on society and individuals. But if viewed as a reasonable definition of mass communication systems, it is also clear that this definition does not capture the full scope of objects that media critics traditionally address. For several decades, scholars have taken issue with the appropriateness of *mass communication* as a term designating the study of the influence of communication technology. At least three issues are noteworthy.

First, the pervasiveness of mass communication systems has been questioned. Gumpert (1970) viewed mass communication as "public" discourse designed for a "large, homogeneous, and anonymous" audience consisting of "a great number of isolated individuals who are not known to each other or by the communicator" (pp. 281–282). He argued that this view of "mass communication is not quite accurate," because the "theories" derived from it "provide only a partial and incomplete explanation of media process and impact" (p. 280). Isolating massive changes in the nature of media systems, Gumpert reported a "trend toward minicomm" or "specific select audiences" (p. 286). In this view, several media—such as magazines, radio, television, and newspapers—increasingly are shifting from targeting a "mass audience" to targeting highly specialized groups. Gumpert concluded, "The 'traditional' concept of mass communication no longer describes 'the way it really is' " (p. 285; see also Gumpert, 1975).

Second, communication technologies are undergoing changes that confound conceptions of them as unidirectional or one-way transmissions. The nature of communication technologies themselves, particularly the rise of interactive media systems, contradict the concept of the passive and unresponsive audience (e.g., Komatsuzaki, 1981; Nilles, 1986).

Finally, more recent findings (e.g., Liebes, 1988) suggest that the mass communication audience is, in fact, composed of multiple audiences, diverse and culturally based, whose responses are distinctive, even unpredictable.

Accordingly, the value of a mass communication orientation has been dramatically challenged (e.g., Cathcart & Gumpert, 1983; Gumpert & Cathcart, 1985). Even McQuail (1987) has noted that the nature of the mass communication audience is shifting from a broad-based "public" to "audiences based on the appeal of certain kinds of content, stars, presenters, authors, etc. At the same time, this also implies, a reduction in the degree and strength of connection between audiences and organized sources. Ties are less likely to be normative and the control of media over their audiences is also lower" (p. 247).

The shift from the mass to diverse audiences occurred, McQuail (1987) has argued, because " 'the audience' is becoming more fragmented in social and spatial composition" and because the "distribution of media content is becoming more separated from production" (p. 247). Although McQuail does not draw the conclusion, his analysis could be a warrant for questioning the reliability and validity of *mass communication* as an enduring theoretical and research orientation for studying communication technologies.

In our view, analyses of media are misdirected if the object of study is mass communication. We believe media should replace mass communication as the more appropriate object of study, a substitution that is significant in terms of what is examined and how it is examined. In this regard, a variety of concepts and terms have been used to analyze media. For example, Gumpert (1970) has argued that the alternative to the study of mass communication should be the study of "mini-comm." He later argued that the alternative should be "uni-comm" (Gumpert, 1975). Almost 10 years later, Cathcart and Gumpert (1983) argued for the study of *mediated interpersonal communication* as an alternative to the study of mass communication. In any event, a variety of approaches exist for scholars seeking an alternative to the study of mass communication. Many of these alternative concerns coalesce around the term *mediated communication*.

Mediated Communication
as an Object of Study and Analysis

As an appropriate object of study, mediated communication is any system of symbol-using in which communication channels function as intervening variables altering message intention. In *Mediated Communication: A Social Action Approach*, Anderson and Meyer (1988) have defined *mediated communication* as the product of two quasi-independent systems of production and reception. In the production system of interlinked media industries, a community of practitioners produces a commodity content for its own end, or as a byproduct in the manufacture of advertising opportunities or commodity audiences. In the reception system, attendance is an intentional performance in its own right, and content is interpreted through its accommodation in the methods and practices of everyday life. (p. 47)

Although useful, this view of *mediated communication* is tied to media industries, a formulation that Anderson and Meyer (1988, p. 50, n. 9) have explicitly acknowledged excludes several other forms of mediated communication. Additionally, this view of *mediated communication* is linked to the intentions and decisions of media industries rather than to the unique formatting or production systems of each technology. Finally, although offering a conception that emphasizes diverse audience reactions, the approach bypasses any consideration of the cognitive structures that adhere in communication technologies. Ultimately, Anderson and Meyer view *mediated communication* in terms of its origins (e.g., media industries) and effects (e.g., interpretive everyday life frameworks), but they do little to identify the intrinsic features of mediated communication.

We see problems in a conception of mediated communication that focuses solely on media industries and bypasses any consideration of the cognitive consequences of technology. At the same time, we believe that *mediated communication* is an appropriate point of departure for isolating the object of study for critical analyses of media systems. Toward this end, we identify six intrinsic features of mediated communication.

First, *mediated communication* emphasizes the form more strongly than the content of messages. The issue here is not to repeat the form–content debate of the early 1920s (e.g., O'Neill, 1923; Sanford, 1922; Wichelns, 1923; Winans, 1923); rather, it calls attention to the forms and mechanisms organizing the relationships among communicators. In this context, Miller (1986) has concluded, for example, that media systems have shifted interpersonal relationships from an "individual" orientation to one in which "other people" are viewed "as undifferentiated role occupants" or "simplistic cultural and sociological caricatures" (p. 136).

Second, *mediated communication* calls attention to the technologies that determine, in part, the meanings attributed to the content of media messages. Elaborating this idea, Cathcart (1987) has maintained,

To claim that the message resides primarily in the narrative located in the verbal text is to miss the actual message, i.e., the one the audience receives and decodes. To ignore the interdiction of media codes and to concentrate on a verbal text, or even a pictorial text, as though verbal or pictorial content alone is what the audience receives and processes, is to limit our analysis to a narrow and often unrealistic rhetorical act that has little or nothing to do with persuasion and the social construction of reality as it functions in contemporary American society. (p. 6)

Cathcart's analysis has been verified under more specific and controlled conditions. For example, Greenfield (1993) compared "the effects of an audio (radio) vs. an audio-visual (television) medium" (p. 4) and found that television and radio present different "opportunities to construct particular kinds of representations" and ultimately stimulate "different kinds of representational processes" that involve different "metacognitive levels of awareness" (p. 3). Accordingly, Greenfield has maintained,

> Cognitive processes—the basic processes by which we take in, transform, remember, create, and communicate information—are universal. But a culture has the power to selectively encourage some cognitive processes, while letting others stay in a relatively undeveloped state. As shared symbol systems, media are potent cultural tools for the selective sculpting of profiles of cognitive processes. A medium is not simply an information channel; as a particular mode of representation, it is also a potential influence on information processing. (p. 2)

Greenfield has concluded, "While individuals respond to and even create media, mass media are also cultural tools. They are both a shared cultural product and a shared cultural representation. To their audience, including children, media not only present culturally relevant content, they also present models and opportunities for particular representational processes" (p. 2).

Third, *mediated communication* highlights the distinctive language of each medium. The formal or formatting technology of each medium generates a distinct language. Cathcart (1987) has explicitly argued: "The mass media, of communication, which are both channel and transmitter of most contemporary rhetoric, cannot be passed off as simple neutral carriers of communication like the air molecules that carry the sound waves of the human voice" (p. 3). Pointing to the works of Ong (1982), Innis (1951), and McLuhan (1962), Cathcart has maintained that each medium "produces a language with its own grammar and syntax" (p. 3). In *Creating Media Culture*, Snow (1983) has formally identified the grammatical features, such as syntax and inflection, of different media systems such as newspapers, novels, magazines, radio, television, and film as well as the sociocultural consequences of these communication technologies. In this regard, Gardner's (1983) seven frames of reference and their corresponding types of intelligence parallel the kinds of information generated by media systems.

In a fuller statement of these views, we can conclude that every medium can be conceived of as functioning as a distinctive and coherent language or logic and cognitive system. The human nervous and cerebral systems literally react differently to the input of each medium. Insofar as the human nervous and cerebral systems are concerned, these different media create different kinds of information, which generate different cognitive understandings. As cognitive understandings vary from one medium to the next, different cultural and moral norms are invoked that govern why certain actions are preferred and taken rather than others.

Fourth, *mediated communication* emphasizes that the language of each medium affects perception and apprehension. Cathcart (1987) has appropriately asked: "What would the critic look for in the media dimension?" (p. 11). His response to this question draws attention to the production processes of each medium: "Just as spoken and written symbols make us attend to certain ideas and not others, so do media systems. Just as words and their ordering concentrate our thoughts, camera shots, cuts, dissolves and fades require us to attend to the logic of the lens and to ignore that which is outside the frame" (p. 11). In this view, the production components of a communication medium are selective channels that convey only certain types of information about events, not all that is important about them (Berg, 1972). As a result, only certain human sensory systems are stimulated. Additionally, the production channels of a medium are the most immediate context of message content, thereby generating a kind of information as important to apprehension as the information generated by content (see, e.g., Chesebro, 1984, 1989).

Employing more precise procedures, Pfau (1990) has explored the degree to which "media" function "as a distinct variable in social influence" (p. 195), focusing specifically on "whether television, as a result of unique channel characteristics, is more similar to interpersonal communication than to public address, print, and radio in the manner that it exercises influence" (p. 199). Pfau established "treatment conditions" in which subjects were exposed to "messages" that "remained constant across all communication modalities," with only the "intrinsic channel features" of each medium able to exert influence (p. 200).

All messages were constructed so as to match as closely as possible total length, language intensity, and comprehensibility. Word counts and a contingency index (Becker, Bavelas, & Braden, 1961) were used to insure similarity in comprehensibility and length of messages. Care was taken to insure that all messages employed similar verb tense, modifiers, and metaphors (Burgoon, Cohen, Miller, & Montgomery, 1978).

Two four-item, seven-point Likert-type scales were created to assess receivers' perception of message content. One four-item scale evaluated the perceived quality of the information presented in the messages. A second four-item scale assessed the perceived strength of the case made on behalf of the product, candidate, and/or cause. Factor analyses were computed and supported the internal consistency of each of the two content dimensions. (Pfau, 1990, p. 201)

We may wonder if the content of any two messages can ever be equivalent if conveyed through two different media systems. However, Pfau's method suggests that a variety of procedures can be employed to identify content differences and adjust message contents to minimize their differences. We might want to see Pfau introduce other measures to determine that all content differences have been minimized, but based on the procedures he employed, Pfau has concluded that "television, like interpersonal communication, elevates person variables in the process of influence" in commercial, political, and social action persuasive messages significantly more frequently than radio, print, and public address do (p. 209).

The methods used by critics also can render similar judgments. In *Eloquence in an Electronic Age: The Transformation of Political Speechmaking*, Jamieson (1988) has isolated the role of media systems in political public address. She has maintained that "what we traditionally knew about eloquence cannot survive" in "this new environment" of electronic media (p. ix). Among other conclusions, Jamieson has maintained that technologies of the electronic age have shifted human communication from an oratorical to an interpersonal style (pp. 165–200).

Thus, regardless of the specific methods applied, it is possible to view production elements as variables that directly affect what people perceive and apprehend. In other words, an idea cannot be examined independently of the production system that gives it form and structure.

In this view, the production techniques of a medium constitute its basic grammar and establish the form and structure that give meaning to a message. For example, in order to create a cinematic message, a frame or single photographic image and the shot or single uninterrupted action of a camera must be used. The frame and the shot function as the basic units of film. At the same time, the frame and shot generate highly selective sensory stimuli because they isolate particular objects in a visuospatial context. In *The Rhetoric of Film*, Harrington (1973) has explored the range of relationships that exist between film production techniques and rhetorical figures. In contrast, the written mode focuses on words or abstract genres that are linearly and sequentially ordered. In *Orality and Literacy: The Technologizing of the Word*, Ong (1982) has explored the communicative implications of print technology.

Fifth, *mediated communication* highlights the cultural systems created and sustained by media systems. Ong (1982) has distinguished, compared, and contrasted the cultural and value systems linked to oral, literate, and electronic communities. In terms of the classical rhetorical canons, for example, an oral culture fosters and reinforces delivery and memory; the literate culture emphasizes style and arrangement; the electronic culture highlights invention. Thus, in Ong's view, media systems constrain human interaction, feature only certain rhetorical activities, and reflect, create, and sustain particular kinds of cultural systems.

Sixth, *mediated communication* focuses attention on the dominant societal metaphor fostered by the media in a culture. In *The Alphabet Effect: The Impact of the Phonetic Alphabet on the Development of Western Civilization*, Logan (1986) has argued that the ways in which communication technologies evolve "has influenced the development of our thought patterns, our social institutions, and our very sense of ourselves" (p. 18). For example, Logan has argued that "because of the neat and uniform way in which information could be organized on the printed page, typography also increased the trend toward uniformity, classification, and analysis" (pp. 193–194) that promoted "self-learning" (p. 195) and became "just about the only medium for the exchange of scientific ideas" (p. 195), which was critical to the development of the Scientific Revolution (pp. 193–209; see also Illich & Sanders, 1988). Additionally, a media orientation apparently can generate reliable and valid findings. Similar societal consequences have been reported when the shift from orality to literacy occurred in the Soviet Union in the late 1920s (Luria, 1976), in medieval England (Clanchy, 1979), and in ancient Greece (Havelock, 1963; see also Enos, 1990). All of this is a way of saying that a medium can be a dominant metaphor in a social system.

As the electronic age unfolds, perhaps to displace literacy as we have understood it, it also may create its own cultural metaphor. For example, Gozzi and Haynes (1991) anticipate that the dominant metaphor of the electronic age will be one of "empathy-at-a-distance" (p. 9) or "ironic empathy" (p. 32). They have initially reported that "the abilities of electric media to simulate people, experiences, and realities are growing," which is "changing" the "nature of knowledge" (pp. 24–25). Focusing on "models" such as "the ironic Johnny Carson, the martyred John Lennon, the humorous nerd Woody Allen, the poet Bob Dylan, the tough Humphrey Bogart, [and] the clear-seeing American Girl," they conclude that "these electric heroes and heroines" provide "one guide to wisdom in the new electric epistemology," a wisdom characterized by a "sense of ironic empathy, compassionate detachment, uninvolved involvement, [and] serious humor" (p. 32). At this moment, we are cautious, and we have yet to be convinced that "ironic empathy" will be the dominant metaphor of the electronic media age. However, we share Gozzi and Haynes's conviction that the electronic media are capable of generating a dominant social metaphor.

By way of definition, we have suggested that six propositions define the parameters of mediated communication. To repeat, mediated communication emphasizes: (1) form more strongly than the content of messages; (2) technologies that influence the meanings attributed to the content of media messages; (3) the distinctive language of each medium; (4) how the language of each medium affects perception and apprehension; (5) the cultural systems created and sustained by media systems; and (6) the dominant social metaphor created by the media that constitute a social system.

This conception of mediated communication is by no means exhaustive or unique. McQuail (1987) has provided a convenient survey of alternative approaches. In addition, three of McQuail's conclusions regarding our approach are apt: (1) "New media and new and expanded uses of communication technology are being widely advocated on the basis of an implied theory of media technology determinism, which is also often a normative theory, giving positive weighing to the maximization of communication possibilities, especially in interactive forms" (p. 315); (2) the approach outlined here is in agreement with McQuail's view that "social progress is assumed to follow and be caused by the expansion of communication of all kinds" (p. 315); and (3) "One of the tasks for the normative branch of a communication science will be to formulate such propositions clearly and provide a framework for putting such theories to the test, under the conditions of information societies which do actually emerge" (p. 315).

If we conclude that mediated communication systems are, indeed, the appropriate object of analysis for a critic, the question turns to how a critical analysis is undertaken. Critical analyses are admittedly complex forms. We have found it most useful to think of criticism as a special type or kind of discourse. Cast in this way, we can initially ask: *What are some of the outstanding characteristics of criticism as a form of discourse?*

| THE NATURE OF CRITICISM

Criticism is a practice, but it also constitutes a body of guidelines, techniques, and applied illustrations that can be used to analyze communication technologies as symbolic and cognitive systems. In this sense, the analysis of media ultimately stems from our understanding of what criticism is. Accordingly, the analysis of communication technologies can be understood as one form or type of discourse derived from a larger genre identified as *criticism*. Given these relationships between the analysis of communication technologies and criticism, we explore conceptions of criticism itself.

In common parlance, the word *criticism* can refer to judgments that are predominantly negative, but the word also can identify evaluation designed to promote understanding of the objects under review. In this context, *Webster's* (1986) has aptly observed that *criticism* is frequently associated with "faultfinding disapproval and objection" as well as "the art of evaluating or analyzing with knowledge and propriety works of art or literature" or "similar considerations of other than literary matters (as moral values or the soundness of scientific hypotheses and procedures)" (p. 539).

Among rhetorical critics, a host of definitions of criticism can be found. We find Campbell's (1979) conception particularly useful, because it identifies the primary objective as well as the means that distinguish criticism from other forms of discourse. Campbell has argued that criticism is epideictic in end and deliberative and forensic in means (pp. 4–13). That is, the intent, objective, and purpose of criticism is to praise/dispraise, judge, or evaluate (see, e.g., Jasinski, 1992, especially p. 198). The means of doing so are deliberative and forensic, which is to say that to sustain a claim of praise or dispraise, a critic must be able to specify what should be (the deliberative) as well as offer reasons and evidence for the claim (the forensic). Thus, critics do more than state their own preferences or tastes; critics offer reasons and evidence for their claims and articulate the implications of their judgments.

Critical discourse has also been characterized as a process of describing, interpreting, and evaluating communicative acts. Brock and Scott (1980) have noted, for example, that the function of the critic is to

> indicate, to point out, to draw attention of others to the phenomenon. . . . In this respect, his purpose is descriptive. With more or less awareness of the implications of his activity, the critic endows with meaning the phenomenon to which he attends. . . . In taking responsibility for his shapings, the critic's purpose becomes interpretive. Finally, the critic judges. . . .
> The primary purposes of rhetorical criticism are to describe, to interpret, and to evaluate. These purposes tend to merge into one another. One purpose prepares for the next; the one that follows reflects back on the one that has been explicated. (pp. 18–19)[*]

---

[*]An alternative to this conception of the "functions" of criticism, in which description, interpretation, and evaluation of criticism are identified as "dimensions" of criticism "as a form of discourse," may be found in Brock, Scott, and Chesebro (1990, pp. 10–22, especially pp. 15–16).

In some cases, critics have even sought to organize their critical analyses into these three major parts. Hence, some critical analyses carry subheads implying that the different parts of the essay progressively and respectively describe, interpret, and evaluate.

However, it is not always easy to discern which statements by critics are descriptive, interpretative, and evaluative. What is taken as purely descriptive will vary. Any "descriptive statement" is a highly selective characterization of an ongoing process that assumes what is most important about the process (a form of interpretation) and also implies a judgment (a form of evaluation) regarding saliency and potency. Because critical discourse increasingly addresses multicultural audiences, distinctions among description, interpretation, and evaluation have blurred significantly. Hence, there may be some senses in which description, interpretation, and evaluation identify interrelated dimensions of critical discourse, but much remains to be done if the description–interpretation–evaluation framework is to define the objectives of criticism or to characterize critical discourse.

Criticism has also been defined in terms of its outcomes or consequences. The outcomes of the critical endeavor can be "to support a speaker's message; to clarify the message so that others will respond as the critic does; to give a message greater significance by increasing the number of people who are aware of it; or to deny the validity of the message" (Chesebro & Hamsher, 1973, p. 282). The outcomes or consequences of the critical effort may not, of course, be as intended or anticipated, and they may not be within the control of the critic.

Finally, many schemes for categorizing criticism have been developed. A few are illustrative. For example, some have distinguished literary and rhetorical criticisms. Wichelns (1925/1966) has suggested that rhetorical criticism deals with "immediate effect" (p. 39) while literary criticism deals with "the permanent values of wisdom and of eloquence, of thought and of beauty" (p. 41). Despite how frequently Wichelns's distinction has been used, it remains unclear why "immediate effect" does not, cannot, and should not deal with wisdom, eloquence, thought, and beauty. Similarly, Campbell (1974, pp. 9–14) has proposed that academic and popular criticism can be meaningfully distinguished. In her view, academic criticism is enduring if it contributes to the body of rhetorical and communication theory, while popular criticism is ephemeral if it illuminates an ongoing transaction in society (see also Foss, 1989, p. 6). Yet ephemeral criticism has stimulated enduring criticism and functioned as a foundation for theoretical explorations and modifications, and criticism is most likely to be enduring because it has aptly captured and reflected the nature of ongoing and pragmatic communicative exchanges in everyday life.

We remain convinced that criticism is epideictic in end and deliberative and forensic in means. We are less convinced that criticism is usefully understood as a descriptive–interpretative–evaluative scheme, in terms of predetermined outcomes or consequences, or by establishing formal categories designed to distinguish forms or types of criticism. We would rather examine examples of critical discourse, identify their common features, and compare and contrast these features with other kinds of discourse. Through such an inductive and comparative process, a body of discourse does emerge that is appropriately identified as *criticism*. We have identified 11 key features here.

1. *Criticism is a form of extensional discourse.* Wood (1976) has reported that interactions can either *extend* the content or semantic meanings of a communicative act (pp. 49–51) or *expand* the existing structure of an idea with qualifiers and responses that elaborate and perfect the original thought (pp. 46–47). Criticism extends the meaning of a communicative act. Criticism is a reaction to a prior communicative act, but a critical reaction goes beyond summarizing the meaning of the original communicative act. Criticism introduces new associations or attributes a new meaning to the original communicative act. As Campbell (1979) has stated the case, "The critical act is a cognitive act; it is designed to make overt what has been hidden" (p. 7).

2. *Criticism is epideictic.* In its most essential form, criticism praises and/or dispraises. In the vernacular, criticism is predominantly associated with dispraise, hence *Webster's* (1986) definition of it as "faultfinding disapproval and objection" (p. 539). Yet *Webster's* also attributes more universal forms of evaluation to criticism, when criticism is defined as "the art of evaluating or analyzing with knowledge and propriety works of art or literature" as well as "similar considerations of other than literary matters (as moral values or the soundness of scientific hypotheses and procedures)" (p. 539). The attention *Webster's* devotes to the epideictic nature of criticism does not depart in significant ways from the conception offered by critics themselves. For example, Black (1965) has argued that critics are engaged in "two vitally important activities": to "see a thing clearly and to record what they have seen precisely," and "to judge the thing justly" (p. 4).

Although criticism necessarily posits a judgment of others' communicative activities, we live in a society in which such evaluations of others often are discouraged, and they can be seen as impolite or antisocial. The growing tolerance for the activities of others and the decision to grant others a wider range of self-determination is generally consistent with an effort to enhance the quality of human symbol using. But this trend is not universal. The symbolic behavior of one group can and frequently does deny that of others. For example, as Foss (1979) demonstrated, the Equal

Rights Amendment generated two diametrically opposed symbolic orientations that revealed "little common ground on which traditional argu-
mentation can occur" (p. 288). Likewise, the symbolic constructions of
realities of "right to life" and "pro-choice" advocates have been cast as
unbridgeable and incompatible, and the conflict has generated a bitter
"battle" that both sides apparently view as decisive in determining what
the quality of human life should be.

These symbolic conflicts reflect the fact that every human being must
respond to the symbolic activities of others, responses that can be essential
for human survival. Accordingly, Burke (1935/1965) proclaimed, "All
living things are critics," reasoning "that all living organisms interpret
many of the signs about them" (p. 5). He illustrates the essential nature
of critical judgments: "A trout, having snatched at a hook but having had
the good luck to escape with a rip in his jaw, may even show by his wiliness
thereafter that he can revise his critical appraisals" (p. 5). In fact, Burke
has concluded that the critical impulse has increasingly become a complex, multilayered symbolic screen defining all human interactions:
"Though all organisms are critics in the sense that they interpret the signs
about them, the experimental, speculative technique made available by
speech would seem to single out the human species as the only one
possessing an equipment for going beyond the criticism of experience to
a criticism of criticism" (p. 6). Burke has concluded,

> We not only interpret the character of events (manifesting in our
> responses all the graduations of fear, apprehension, misgiving, expec-
> tation, assurance for which there are rough behavioristic counterparts
> in animals)—we may also interpret our interpretations. The need to
> evaluate, and then re-evaluate, is a way in which value priorities can
> constantly be assessed as new circumstances are encountered. (p. 6)

Although it is a normal and essential feature of the human condition,
the need to evaluate must be held in check. When evaluation functions
as the only frame of reference, the tendency to be judgmental can create
closure. A judgmental orientation may prevent a critic from considering
how one communicative act can mean different things to different audiences. Consequently, the natural tendency toward the judgmental can
preclude social cohesion and unity.

Despite the limitations of criticism as evaluation, the techniques
essential to develop a critical posture—the search for alternatives (the
deliberative) and the necessity to provide good reasons and evidence (the
forensic)—act as built-in restrictions on a mindless use of evaluation.
Whatever the perils of evaluation, the need to develop such skills decisively outweighs this concern. Hacker and Coste (1992) vividly illustrate

the overwhelming need for all individuals to develop evaluative skills. They have observed that television news viewers "often challenge newscasts on a surface level and, less often, on a deeper level regarding bias, news producer intentions, and truth claims" (p. 299). They concluded, however, that "there is little evidence to suggest that these viewers resisted the ideological codes of the news. This does not mean that they did not have the ability to resist news ideology. Instead, it seems to be unusual for viewers to do so. A major obstacle to news viewer resistance seems to be lack of alternative significations from which to draw when formulating semiotic narratives about news events" (p. 299). As communication technologies and media systems increase exponentially, the need to evaluate critically has increased dramatically.

3. *Criticism is deliberative*. At every turn, criticism is deeply involved in politics. Insofar as politics involves power relationships, persuading others, shifts in hierarchies, and influencing or controlling the destiny of others, criticism is political. Selecting one communicative act for critical review rather than another is itself a political act because the critic implicitly infuses the selected act with greater significance. Additionally, criticism is a political act because the critic praises/dispraises the ideas in the communicative act analyzed. Finally, whether the critic confirms or rejects the existing social order, assessments of social planning and future actions are profoundly political activities. As Black (1965) aptly summarized the issue, critics differ from scientists because they seek to make their criticism "a force in society" (p. 5).

4. *Criticism is reason giving*. A statement such as "I hated it," without reasons and evidence, is only a statement of personal preference. Although such statements are informative, they are not criticism. For example, Andrews (1983) has maintained, "A critic combines knowledge with a systematic way of using that knowledge and constantly seeks to refine his or her practice of criticism" (p. 5). He concluded that "in the most fundamental sense the critic is an educator" (p. 6), a view shared by Black (1965) when he suggested that "the critic is an educator, and insofar as he fails to educate, he fails his essential office" (p. 6). As a reason-giving and educational activity, criticism should be verifiable. A test of good criticism is whether the critic's claims can be confirmed with reasonable research and can be said to be more true than false under the conditions specified by the critic.

5. *Criticism is self-reflexive and ideological*. Minimally, most critics seek to offer a "fair" reading of the communicative act they examine. Yet even the fairest reading reflects the critic's personal understandings. The critic selects what he or she believes are the "salient" features and thereby constructs an autobiographical conception of the communicative act. Even the most catholic of critics, encouraged by the best of intentions,

can provide only a partial reading. Such a reading will feature the vested interests of some at the expense of others. The recognition that power differentials exist in society establishes a foundation for concluding that all criticism is necessarily ideological. Wander (1983) has argued, "Criticism takes an ideological turn when it recognizes the existence of powerful vested interests benefiting from and consistently urging policies and technology that threaten life on this planet, when it realizes that we search for alternatives. The situation is being constructed; it will not be averted either by ignoring it or placing it beyond our province" (p. 18). Hence, Wander has maintained, "An ideological turn in modern criticism reflects the existence of crisis, acknowledges the influence of established interests and the reality of alternative world-views, and commends rhetorical analyses not only of the actions implied but also of the interests represented" (p. 18). In this sense, the evaluative dimension of criticism necessarily involves adopting some kind of an ideological posture.

Yet the ideological nature of criticism should not discourage critics, nor does it mean that critics should not attempt to offer the most complete description, interpretation, and evaluation of a communicative act. Recognizing the ideological nature of criticism means that what a critic praises/dispraises will be limited in its universality. Thus, acknowledging the ideological nature of criticism should humble critics who otherwise might perceive themselves as expert judges with the final word on the value of a communicative act.

6. *Criticism is persuasion.* It is persuasive because it suggests an alternative view of a communicative act, regardless of whether the alternative view challenges the validity of the idea. The attempt to reframe a communicative act is an effort to change attitudes toward it, an effort that falls within the province of persuasion. Additionally, the decision to interpret or classify a communicative act as one social activity rather than another and the decision to praise/dispraise a communicative act with reasons and evidence are classic examples of the most traditional persuasive strategies. In doing criticism, critics necessarily employ persuasive techniques. Accordingly, when they offer critical analyses, critics enter the rhetorical arena, and they function as advocates who should anticipate rejoinders and challenges.

7. *Criticism assesses the effectiveness of communicative acts.* Rhetorical critics have traditionally focused on the strategic choices made by a speaker to achieve his or her goals and offered judgments about the degree to which a speaker achieves these goals. To assess such effectiveness limits the potential role of criticism. To the extent that a critic assesses effects in terms of a speaker's intent, the critic becomes, as Wander (1983) has suggested, a "public relations consultant" for the speaker (p. 9).

More appropriately, the critic should be examining the effect of a speech on society. As Thonssen, Baird, and Braden (1970) noted over 25 years ago, "Rhetorical criticism helps to interpret the function of oral communication in society" (p. 24). More recently, as Bormann (1972) has demonstrated, the ideas contained within a single speech can "chain out" from the immediate audience to an ever-increasing number of audiences and "into the mass media and, in turn spread out across larger publics" (p. 398). In this way, the ideas of the single speaker can "catch up larger groups of people in a symbolic reality" (p. 398). Thus, the social consequences of a communicative act are appropriate grounds for the critic's assessment.

Additionally, the critic also might focus on a communicative act that has not made a difference, but which, in the critic's view, should reach wider audiences. The critic can focus on communicative acts that have actually generated significant responses or ones that should have been, but were not, significant in this sense.

Although an "effects orientation" is particularly important to understand communication and criticism, its advantages and limitations are evident. To examine the consequences of a communicative act reveals how symbol using functions instrumentally in society as well as how specific strategies work. At the same time, the decision to focus on consequences may shift attention away from the moral stance embedded in a symbolic act, promote the notion that a single act "causes" the consequences that follow (i.e., a single cause-to-effect model), and suggest that a social engineering model adequately explains human communication.

8. *Criticism involves an appreciation of form.* Aesthetic in its emphasis, criticism assesses the artistic merit of a communicative act. Traditionally, criticism with this purpose focuses on form rather than content. Aesthetic assessments typically assume that any single communicative act can be compared to and contrasted with other communicative acts. Comparison relies on classification. Such classifications generally are identified as genres. Relying on the earlier work by Black (1965), Campbell and Jamieson (1978, p. 14) have identified four of the assumptions of the generic approach to criticism: (1) "there is a limited number of situations in which a rhetor can find himself"; (2) "there is a limited number of ways in which a rhetor can and will respond rhetorically to any given situational type"; (3) "the recurrence of a given situational type through history will provide the critic with information on the rhetorical responses available in that situation"; and (4) "although we can expect congregations of rhetorical discourses to form at distinct points along the scale, these points will be more or less arbitrary."

When communicative acts are ordered into their respective genres, the distinctive features of a genre can be identified or described. Each act can then be viewed as more or less artistic in terms of its execution.

Additionally, the distinctive characteristics of each category can also be rank-ordered in terms of their relative importance.

Aesthetic assessments allow critics to fashion links among communicative acts regardless of their immediate audiences and historical contexts and thereby offer judgments that are unlikely to be evident to those caught up in an immediate event. Generic assessments therefore are likely to generate special insights, because this mode of assessment necessarily provides a cross-cultural perspective on a communicative act, which is thus viewed in terms of alternative sociocultural systems. At the same time, genres can be affected by changing circumstances. Accordingly, with time, genres may blur, and some have suggested that the mass media have had this effect on generic formulations (see, e.g., Avery & McCain, 1986), while others have maintained that mass media systems have so dramatically changed human communication that an entirely new generic framework is warranted (see, e.g., Cathcart & Gumpert, 1983). Additionally, aesthetic judgments may also be limited because they slight the specific circumstances and practical issues that shaped how a specific communicative act was formulated and executed.

9. *Criticism constitutes an exploration of the applied and theoretical.* Particularly within a narrowly conceived view of the discipline of communication, signified by a growing body of criticism in the journals of professional education associations, criticism has been used to achieve one of two ends. Criticism can contribute to theories of communication. Such criticism is valued for its contribution to the discipline and has been characterized as enduring. Criticism also has been designed to explain ongoing communication transactions. Such criticism is valued for its ability to reveal and resolve immediate and practical societal issues, but it has been called ephemeral. These dual functions—the theoretical and the applied—have been examined in some detail by Campbell (1974) and Foss (1989, p. 6). These functions have been attributed to criticism for almost half a century. Writing in 1948 soley about oral communication, Thonssen and Baird maintained that criticism

> helps to clarify and define the theoretical basis of public address. It does so without proposing to teach the speaker how to manage his [or her] art. . . . [C]riticism helps to reveal the operation of theory in practice, thus clarifying its meaning and perhaps in some instances even formulating new theory. . . . Rhetorical criticism [also] helps to interpret the function of oral communication in society. It serves as an effective link between the theory of public address and the outside world. . . . [C]riticism traces the major steps in oral communication straight through to the *effect,* immediate or delayed, of the spoken discourse upon society. (p. 21)

10. *Criticism is a form of self-exploration and self-expression.* Drawing attention to the idiosyncratic features of criticism, Black (1965) has held that "Criticism is that which critics do" (p. 4). Criticism can be designed by critics to articulate their views. Although it may serve other ends as well, such criticism is a self-expression of the critic. Of course, all criticism reflects the unique orientation and position of a critic, yet some criticism may be generated to reveal a critic's personal commitment to a public audience. Indeed, such criticism may be explicitly characterized by the stance: "It's time to take a stand even when no one else will." In some cases, revealing one's personal perspective can serve social ends. A specific example illustrates how this personal stance can serve larger societal objectives. Writing for the predominantly African-American readership of the *Journal of Black Studies* in 1970, at the peak of the Black Power movement, Scott (1970) adopted such a personal stance in "Rhetoric, Black Power, and Baldwin's 'Another Country' ":

> This piece is rhetorical criticism. That statement may doom the writer at the outset to an absurd posture. Criticism suggests detachment, but the subject is one that mocks such an attitude. Further, striking the critical stance may be innately arrogant and, surely in the searing gospels of the American ghettos even more than in those of yore, does a haughty spirit go before destruction. . . . Description and evaluation both depend on understanding. But understanding is never absolute. It reveals a point of view. . . . A simple statement of position, then, may be vital. In the present context, for example, my reference points would include: liberal, white, male, over forty. Such a statement may be vital but is never sufficient and is always misleading in its simplicity. For every human position carries with it a sense of how-I-got-there and where-I'm going. (pp. 21–22)

Explicitly noting that the "urge to tell it like it is" may mean "to tell it as it seems to me" and also noting that "from one point of view, the obvious disadvantage becomes an advantage," Scott adopts the personal stance in order to provide "feedback" from a "white to black" perspective (pp. 22–23).

Criticism cast as self-exploration or self-expression reflects our earlier observation that all criticism is necessarily self-reflexive and ideological. At the same time, recognizing the personal nature of criticism does not mean that criticism is merely self-indulgent. Gardner (1983, pp. 237–276) has argued that understanding the "sense of self" is a specialized type of intelligence that all human beings possess and need to exhibit. He also has maintained that some people are more effective than others in using their "personal intelligences" (i.e., the ability to understand their intrapersonal and interpersonal relationships). Some individuals, Gardner has observed, have a greater "core capacity" to gain "*access to one's own feeling life*" and "*to notice and make distinctions among other individuals* and, in particular, among their moods, temperaments, motivations, and intentions" (p. 239). The exploration of the self and the ability to express the self can be appreciated as a form of intelligence that can be more or less appropriate in criticism depending on the ends it serves.

We are left with several conclusions. All criticism is self-reflexive. All criticism reflects, in various degrees, self-indulgence. As a form of intelligence, self-exploration can generate insights. Such elements in criticism can be appreciated, depending on the degree to which they promote insightful assessments of communication.

11. *Criticism is entertainment.* Altheide and Snow (1979) have explained this universal view of entertainment, discussing the "entertainment perspective" as involving any "behavior that is extraordinary" or "outside the expected limits of routine behavior" (p. 20). They explain: "It is not enough to say that something is innately funny or that some people are simply more entertaining than others. Entertainment consists of a general perspective, a way of 'seeing' and making sense of behavior. Analytically, the entertainment perspective consists of a set of norms or criteria for presenting and evaluating behavior" (p. 19). Thus, Altheide and Snow concluded,

> Entertainment also may be understood as a perspective consisting of norms that are quite different from those applied to routine everyday life. What differentiates entertainment criteria from the mundane is that, excluding content, one is the opposite of the other. We might visualize a continuum with the mundane at one pole and entertainment at the other. The content along this continuum may not change, but how it is defined certainly does change. The entertainment perspective seems to involve behavior that is extraordinary, highly skilled (talent), allows for vicarious audience involvement, and is enjoyable in a fun or play sense. The extraordinary characteristic of entertainment means that it is behavior outside the expected limits of routine behavior. (p. 20)

In other words, criticism is entertainment because it offers new insights to intrigue or educate audiences.

Even in its more common meaning, criticism *should* be entertaining. If criticism is to praise or dispraise, it should at least equal, if not exceed, the ideational and persuasive quality of what is criticized. Criticism works against itself if it fails to excite while praising other communicative acts for their instigating and fomenting features. In this view, criticism should be able to equal the rhetorical effectiveness of the acts it assesses. If criticism is truly extensional, if it renders judgments, plays a political role, and offers reasons for its conclusions as it reflects the ideology of the critic, then it should enact the standards it uses to evaluate. In all, criticism should be a model of effective communication, creating new insights and understandings, intriguing and entertaining audiences.

Given these views, it is appropriate to ask if criticism possesses any unique characteristics when it is applied to the analysis of communication technologies. Such a question turns our attention to an explication of the purposes of media criticism.

| PURPOSES OF MEDIA CRITICISM

Despite its widespread occurrence in popular and academic environments, conceptions and definitions of the media analysis process lack precision and frequently are at odds. Characterizing journalism and news media management critics, in *Criticizing the Media: Empirical Approaches*, Lemert (1989) reported that "no consensus exists" on "the values, standards, and procedures to be used in critical analysis" (p. 10). Similarly, in *Media Criticism: Journeys in Interpretation*, Scodari and Thorpe (1992) concluded that, "Depending on context, purpose and/or circumstance, the definition of media criticism can vary" (p. 3). A few illustrations, with some critical notes, of efforts to resolve this ambiguity are appropriate, because they reveal the difficulties of the definitional task at hand.

Although they recognize that the "definition of media criticism can vary," Scodari and Thorpe bypass the practical issue of defining what media criticism actually is. Shifting from a descriptive to prescriptive framework, Scodari and Thorpe (1992) concern themselves only with "legitimate media criticism," and they predetermine what media criticism *should* and *should not* be:

> Simply stated, legitimate media criticism involves the intellectual, subjective analysis and/or evaluation of media artifacts, policies, technologies, and/or institutions by "disinterested" persons who do not stand to personally profit as a consequence of their specific criticism. Although approached subjectively in order to render a judgment or increase understanding, criticism's claims should be validated through logical, well-supported arguments. Furthermore, as much as criticism may benefit by using "scientific" research data to lend further credibility to the arguments advanced, it is the subjective, interpretive approach and/or the evaluative element that ultimately labels the discourse as criticism. (p. 3)

The goal of the analysis of communication technologies is the explication of the cognitive, behavioral, and motivational dimensions and the consequences of communication media in an effort to enhance the quality of human symbol using. As discussed earlier, these critical efforts are epideictic in end and deliberative and forensic in means. A corpus of such criticism emerges when the principles of criticism are consistently directed toward the analysis of communication technologies. Media criticism is the product of the consistent application of critical principles to media technologies as communication systems. We will use the concept *media criticism* to refer to the corpus of thought and analyses that critics have generated regarding communication technologies during the last 50 years. Within an admittedly pragmatic framework in which the analysis of communication technologies may be conducted for the first time, we think it is useful to isolate some of the specific purposes and objectives that have emerged when communication technologies are described, interpreted, and evaluated.

In one sense, the 11 characteristics of criticism identified earlier in this chapter would reveal how and why the analyses of communication technologies are undertaken. Yet in terms of a more practical sense of what an analysis of a communication technology is likely to be, it also is particularly useful to review some of the analysis-specific objectives that have emerged when critics actually assess communication technologies.

Within the context of communication technologies and media systems, criticism can achieve one or more objectives, and it can vary in emphasis. Some of the objectives will receive more attention than others, depending on the critic, the communicative act, the strategies being examined, the judgment rendered by the critic, and the circumstances influencing the critic. Given our previous discussions of the characteristics of criticism, we list only some of the objectives of media criticism, and we invite the reader to suggest how they might shape assessments of our media world:

First, media criticism can reveal the subtle and unnoticed complex stimuli embedded in media experiences. In other words, media criticism can help us "see more" than we would from a casual viewing. A mediated experience presents a complex set of multilayered stimuli. An analogy helps to reveal how media criticism can reveal the unobtrusive. If we are unaware of the roles and functions of nonverbal communication in face-to-face conversations, we might concentrate on what we hear and be unconscious of the fact that a person's verbal message is not believable because nonverbal communication is interfering with or contradicting what is being said. Once we are aware of the influence nonverbal communication exerts, and pay equal attention to verbal and nonverbal dimensions, face-to-face conversations become more informative, useful, and interesting. Similarly, media criticism can increase our awareness of the total volume of perceived cues or information within a media presentation. The subject in a film, for example, must be shot from a high, standard, or low angle. Although not explicitly identified as an important feature of the scene, camera placement influences how viewers value the subject being filmed. A high-angle shot creates the impression that the person is being "looked down upon and made to appear weak," while a low-angle shot provides a subject with "an air of dominance and of being larger than life" (Harrington, 1973, p. 77). In this case, the camera placement construct (i.e., high, standard, and low angle) employed by the media critic can make us more aware of the influences of camera angle. In such ways, given its focus on the format and production syntax of a communication technology, media criticism can reveal what might otherwise go unnoticed.

Second, media criticism can reveal how media systems affect human cognition. Media criticism reveals the knowledge generated by the formatting and production syntax of a medium, but it also can isolate new knowledge environments created by a medium. For example, Cathcart (1986) has argued that media systems create new communities, "media communities," that link people together who otherwise would not interact face to face. These media communities seldom occupy a common geographic context; media communities exist independent of any specific space or time (see, e.g., Drucker & Gumpert, 1991). Noting that "technological media innovations provide the world to each person without the need for physical proximity," Cathcart (1986) has maintained that the " 'media community' rests upon the notion that interaction can be facilitated by media connection and that the absence of physicality does not hinder relationships" (p. 3). For example, the "world of television football fans" is not linked to any specific geographic region. Viewers from all over the United States jointly "participate" with other football fans, regardless of where a game is played and without direct contact with the other fans. There are, of course, several factors in a game broadcast that orchestrate the common reactions of television football fans, but the larger point here is that a sense of community is not linked to face-to-face interactions or to a common geographic location (see, e.g., Brummett, 1992). Similarly, a media community might exist independent of a shared time reference. People reading and leaving messages on computer bulletin boards can do so at any hour, and in the case of computer bulletin boards, participants may form "friendships" without knowing—or having to know—the demographics (e.g., age, gender, race, etc.) of those with whom they are communicating (see, e.g., Chesebro & Bonsall, 1989, pp. 100–103). In such cases, media criticism isolates social communities and knowledge environments that are solely a function of communication technology.

Third, media criticism can reveal the effects of media systems, particularly their effects on individuals and society. It is difficult to imagine a media critic who can avoid questions of impact or consequences. A media experience is unworthy of analysis unless it has the potential to influence either individuals or society, and the degree of influence itself frequently determines the significance of the media experience. In these senses, questions of effect permeate virtually all media criticisms.

Yet attempts to determine the effect of any media experience have become increasingly complicated, particularly when media systems are involved. Rogers (1986) has noted that

> most people just let the mass media sort of wash over them. Television has become a kind of video wallpaper; a large percentage of Americans passively watch their "least objectional program," absorbing little of the message content. When asked to recall certain salient facts from a TV news broadcast within a few hours of viewing it, few can do so. Many new TV sets and video cassette recorders come with a remote control that makes it easy to switch channels or to "zap" commercials, but only 2 or 3 percent of all viewers do it. (p. 31)

Having considered these possibilities, Rogers has concluded, "So the general picture that emerges of the mass media audience in American is that of passive receivers" (p. 31).

Rogers has accordingly reported that attempts to identify any "direct effects" have been less than successful: "Mass communication ordinarily does not serve as a necessary and sufficient cause of audience effects" (p. 154). "Somewhat greater optimism," Rogers has argued, has been demonstrated by examining the "*indirect effects*" of media systems (p. 156). These indirect effects occur "among and through a nexus of mediating factors and influences. These mediating factors are such that they typically render mass communication a contributory agent, but not the sole cause, in a process of reinforcing the existing conditions" (Klapper, 1960, p. 8; quoted in Rogers, 1986, p. 154).

Given these limitations in any assessment of media effects, Table 2 is a useful conception of the range and consequences of media systems.

In this scheme, it is assumed that any consequences of media systems will be indirect, and any outcome is a product of multiple and converging social forces of which media systems will be only one cause. In addition, it is assumed that a media critic can focus on either how media systems and human beings interact (a process orientation) or the outcomes and consequences generated as a result of a media–human interaction, with particular attention given to the changes in human behavior, actions, and attitudes.

TABLE 2. Range and Consequences of Media Systems

	Process	Consequences
Individual		
Personal relations		
Parasocial relations		
Society		

Table 2 also suggests that a media critic might usefully examine media consequences in terms of particular individuals, the face-to-face interactions defining the interpersonal or social relationships among individuals, the parasocial relationships individuals establish with media personalities (see, e.g., Horton & Wohl, 1956, cited in Gumpert & Cathcart, 1986, pp. 185–206), and social institutions and structures, cultural norms and values, rules and organizational systems, codes, and roles in society (see, e.g., Lindesmith & Strauss, 1956).

Fourth, media criticism can alter the communication process by introducing quality-control components into the media communication system. Media criticism is seldom identified as a component in traditional models of the communication process (see, e.g., Berlo, 1960). Media models typically identify a series of components that account for how and why mediated communication occurs. These components generally include the source (e.g., producers, industries, writers, talent, etc.), channels, messages, contexts, receivers, and even noise (i.e., any factor that interferes with the transmission of a message from a source to receivers). Seldom is media criticism treated as a discrete component in the mediated communication process equal in power to the influence attributed to the source, channel, message, contexts, and receivers. Yet media criticism can be generated often enough and have sufficient impact on audiences, that, for all practical purposes, it can become one of the basic elements of the mediated communication process. In such a situation, criticism is a consistent and formal evaluative dimension within the mediated communication process, a built-in quality-control device within the system. In many respects, in part because of the ever-increasing price of movie tickets, film critics may already have attained such a status. With time, such critics can be perceived as an institutional force within a media industry, exerting the same—if not more—influence as a major production firm. In such cases, media criticism alters the communication process and introduces a quality-control component into the system.

Fifth, media criticism can create counterarguments to the messages generated by media technologies. In some cases, critics are explicit in their objective. For example, in his aptly entitled volume, *Four Arguments for the Elimination of Television*, Mander (1978) has argued that television should be eliminated because it deprives viewers of direct environmental sensations, standardizes and controls experiences, suppresses human creativity, and creates an alien and nonhuman world. Television is unlikely to be eliminated, but Mander's counterarguments apparently stem from a stance designed to challenge televised messages: "I came to the conclusion that like other modern technologies which now surround our lives, advertising, television and most mass media predetermine their own ultimate use and effect. In the end, I became horrified by them, as I observed the aberrations which they inevitably create in the world" (Mander, 1978, p. 13). In such cases, media criticism offers a profound alternative to the symbolic environment created by communication technologies.

Sixth, media criticism can generate individual, rather than social class, reactions to media experiences, energizing the media user in an attempt to shift him or her from an inactive to an active role. That is, media criticism can foster unique and individual reactions to media experiences. In this view, media criticism might contribute to the discourse that makes it impossible to predict individual reactions to media experiences solely in terms of the demographic characteristics of an individual or the groups to which an individual affiliates or belongs. In this sense, media criticism can foster a redefinition of the self.

We would invite readers to determine how and why each of these objectives is important in a human environment dominated by communication technologies. We would also invite readers to consider which of these objectives are more important than others. We are firmly convinced that the most important objective of media criticism is to generate individual, rather than class, reactions to media experiences.

A MODEL FOR THE ANALYSIS OF COMMUNICATION TECHNOLOGIES: DEFINING AND EVALUATING THE TECHNOCULTURAL DRAMA

Any number of methods can be proposed to analyze communication technologies. Each one draws attention to different communication variables. The model proposed here is equally selective. It treats communication technologies as texts that are to be understood in terms of cultural systems. The interaction between technological texts and cultural systems generates a series of technocultural dramas, that constitute the object of study for critics analyzing communication technologies. Given the more precise terminologies introduced at this juncture to describe this model, we begin with an overview of the model to highlight the meanings of concepts such as *technology as text, cultural system*, and *technocultural dramas*.

As we have implied, the concept of *technology as text* shifts attention from the content of a communication technology to the structural features of the technology itself. In greater detail, three meanings are embedded in the notion of *technology as text* (cf. Woolgar, 1991).

First, the concept emphasizes the instrumental functions of a technology. In this sense, technologies are not "simply docile objects with fixed attributes (uses, capabilities, and so on)" (Woolgar, 1991, p. 37). The meaning of the technology or the text of the technology is not revealed by examining the intentions of its designers. They may provide little indication of its instrumental functions. In this context, computer technologies are an appropriate example. Waltz (1982) has noted that, "In many cases the programmer does not know what his program can do until it is run on a computer" (p. 130). Similarly, several computer programs, such as heuristic, planning, backward-chaining, and concept-learning programs, are designed to alter human inputs and to generate "unique results with a minimum of guidance from the user" in which the "outcome of one of these computer runs is unknown" (Chesebro & Bonsall, 1989, pp. 179–180). As Williams (1982) has noted, computers "can qualify as *communication technology*" because "they are capable of taking our messages and giving them back to us or others," but "unlike any other communications device, they are capable of acting upon them in a manner defined by an extension of our own human intelligence" (p. 108). As Miller (1978) has maintained, a technology is different than a tool. Tools, "extend the immediate biological capabilities" of human beings, while by definition technologies lose their original purpose and begin to function independent of the human being (p. 229). Accordingly, after outlining an expert system for "speech composition," Phillips and Erlwein (1988) concluded that "no argument is made that the composition program itself operates in human fashion" (p. 258). Ultimately, the instrumental functions or meaning of the text of a technology is determined by examining the full range and permutations of its production capabilities.

Second, *technology as text* emphasizes its interpretative functions. A technology organizes the environment in which it operates by drawing on selective resources from its environment and affecting selective features of its environment. The invention and mass production of the automobile has affected, for example, how our cities are organized. The technology creates its own "organizational structure, management style, beliefs, and culture" (Woolgar, 1991, pp. 37–38). To function effectively in a technology's environment, practitioners must accommodate or adjust to the technology by establishing functional connections between the technological environment and their practices. Ultimately, the interpretative functions or meanings of the text are determined by identifying the coding system or the organizational form used to construct the "reality" to which a technology can respond.

Third, *technology as text* emphasizes its reflexive functions. Cultural systems theoretically can respond to and use a technology in an infinite number of ways. A cultural system might seek, for example, to avoid any kind of interaction with a technology. However, if a cultural system uses the services or products generated by a technology, that use reduces the range of potential responses, if the full capabilities of the technology are to be realized. For example, directions, time, space, and urban planning are conceived and articulated in terms of the requirements of driving an automobile. In essence, when a technology is used, its reflexive functions constrain the possible range of human responses. Ultimately, the reflexive functions or meanings of a technology are determined by the range of responses that are possible if its outcomes are to be realized.

In these three senses, a technology can be said to possess a text. For the text to be "read," the restrictions created by the technology itself must be accommodated. A technology possesses instrumental, interpretative, and reflexive functions that regulate how human beings can respond to a technology.

At the same time, technologies are a product of cultural systems. As we have noted in Chapter 1, "*Culture, then, consists of standards for deciding what is, standards for deciding what can be, standards for deciding how one feels about it, standards for deciding what to do about it, and standards for deciding how to go about doing it*" (Goodenough, 1971, pp. 22). The cultural standards and norms (or values) of a societal system stimulate the development of and foster the use of technologies. Indeed, a cultural system is also capable of determining how a technology is to be used. As Woolgar (1991) has stated the case, "society plays an important part in deciding which technologies are adopted" (p. 30). He has explained, "there are many instances when devices judged useful and even essential were not taken up or were effectively resisted" (p. 30). In all, "determining the effects of a technology" requires "an understanding of the overall dynamics of society" (Woolgar, 1991, p. 30).

Theoretically, the text of a technology and the norms of a cultural system can each be viewed as deterministic. Technology and culture also might be viewed as dialectically related. On one hand, as a text, a technology can restrict human actions and responses. On the other hand, as the system that generates a technology, a culture can restrict how technologies are or are not used. In practice, the text of a technology and the norms of a culture frequently interact, creating an outcome that reflects both.

The term *technocultural drama* captures the dynamic relationship that exists whenever technology and culture interact to create a system in which they mutually define the nature of the human experience. Any number of explanations might be posited for this technology–culture interaction. For example, Hughes (1983) has speculated that what we refer to as technocultural dramas are a "natural" outgrowth of the human experience because human beings are *"technological"* and can be characterized by a "creative spirit manifesting itself in the building of a human-made world patterned by machines, mega-machines, and systems," that grow from a value system of "order, system and control that [has been purposefully] embedded in machines, devices, processes, and systems" (pp. 3–4).

Although discussions regarding the origins of the technological impulse are interesting, the stages of the dynamic and "corrective" relationship between technology and culture are more salient. Again, several systems could characterize how technology and culture interact over time. For example, using the term "technological drama" (for which we use the term "technocultural drama") to reflect his concern, Pfaffenberger (1992) has argued that "a technological drama is a discourse of technological 'statements' and 'counterstatements,' in which there are three recognizable processes: *technological regularization, technological adjustment,* and *technological reconstitution*" (p. 285). "A technological drama begins," Pfaffenberg has argued, "with technological regularization" in which the designer of a technology "creates, appropriates, or modifies a technological production process" so that the technology's "technical features embody a political aim" or "intention to alter the allocation of power, prestige or wealth in a social formation" (p. 285). This period of technological regularization is followed by a period of technological adjustment in which "the people who lose when a new production process" is introduced begin to "engage in strategies that try to compensate for the loss of self-esteem, social prestige, and social power that the technology has caused" (p. 286). The period of technological adjustment is followed by a period of technological reconstitution in which "impact constituencies try to reverse the implications of a technology through a symbolic inversion process," that frequently leads to "the fabrication of *counterartifacts*, such as the personal computer or 'appropriate technology,' which embody features believed to negate or reverse the political implications of the dominant system" (p. 286). Pfaffenberg concluded that, "I choose the metaphor *drama* rather than *text* to describe these processes," because the processes "draw deeply from a culture's root paradigms, its fundamental and axiomatic propositions about the nature of social life, and in consequence, technological activities bring deeply entrenched moral imperatives into prominence" (p. 286).

Based on such reasoning, the model proposed here is particularly useful when addressing four specific questions, as follows: (1) What kinds of selective information are and are not generated by a communication technology? (2) In terms of knowledge systems, what kinds of knowledge or cognitive structure are consistent with the information generated by each communication technology? That is, cast in terms of individual learning, what can people apprehend, understand, or know using the information conveyed by a given communication medium? (3) What social institutions are created when these communication technologies function as dominant ways of knowing? (4) Treated as symbolic and cognitive systems, what are the positive and negative consequences of each kind of communication technology?

Our model features four interdependent and interrelated dimensions:

1. *Structural analysis:* A critic initially seeks to determine the full range and permutations of the production capabilities of the technology. Several questions are useful: What components of the medium shape perception of its content? What kind of information is conveyed by the medium? What kind of information is ignored or excluded by the medium? How are the production elements of a medium engineered or structured to emphasize and highlight certain information but not other information? The critic ultimately focuses on how the format of each medium selectively records and conveys highly selective information to the user of the communication technology.

2. *Cognitive analysis:* At this stage, a critic determines how the format or technology of a medium shapes human apprehension and subsequently how the communication technology affects how people understand what does and does not exist. Because each medium offers a selective and different conception of the physical and social realities it reports or broadcasts, the critic seeks to determine how this selective conception influences why and how viewers understand phenomena in one way and not others.

Yet, at this juncture, we also need to be particularly conscious of the contribution of postmodern thinkers. Postmodern rhetorical theory and criticism, developed in the mid-1960s in Western Europe, has offered one of the most thorough challenges to traditional American rhetorical theory and criticism. It would be inappropriate to attempt to articulate the entire structure of postmodernism here (see, e.g., Foss, Foss, & Trapp, 1985). We want to draw attention to one postmodern principle however: *Every symbolic act inherently conveys multiple and contradictory meanings* (see Brock, Scott, & Chesebro, 1990, pp. 428–500). We interpret this postmodern principle to mean that different people can interpret the messages of media systems in different ways (see, e.g., Liebes, 1988). Given diverse understandings, there may be times when a critic will find it useful to ask if a range of reactions or responses to the same communication technology exist. The urgency of such an analysis will turn on the kind of issue critics examine, but as part of the basic mode of any analysis of a communication technology, we would urge that such a consideration be made.

3. *Sociological analysis:* Media systems unify and divide people into different social groupings, and each of these social groupings is more or less powerful compared to others. The more powerful social groupings are capable of creating, sustaining, and altering societal processes and institutions that regulate and control other social groupings. At this stage, the critic asks *how* societal processes and institutions are created, sustained, and altered by media systems. Critics may decide to employ a flexible conception of sociology at this juncture, and they may view historical and transnational events and trends, such as the Industrial Revolution and Renaissance, as social institutions.

4. *Evaluative analysis:* In judging the final outcome, a critic might ask, How do the structural, cognitive, and sociological dimensions of a communication technology affect the meaning of the personal life and the social systems in which an individual exists? Designed to enhance the quality of symbol using, the critic ultimately hopes that those reacting to a critical analysis will no longer employ a set of "class reactions" when encountering media formatting systems but react to media systems as unique individuals. For us, the intellectual and pedagogical goal of media criticism is to foster an individual evaluative framework in all those who come in contact with communication technologies.

| CONCLUSION

In this chapter, we examined several features of criticism. We initially isolated some of the changes that have emerged when analyzing communication technologies. We then proposed that mediated communication is the appropriate object of study for the analysis of communication technologies. Pragmatic in its orientation, this chapter also has provided specific guidelines and examples for the analysis of communication technologies. Accordingly, in this chapter, we considered the distinctive features of criticism that emerge when communication technologies are examined, including a particular model for investigating communication technologies.

PART III Media Cultures

Communication technologies have frequently overwhelmed human beings. Shortly after photography was introduced in 1822, some cultures believed a photograph trapped the soul of those photographed. In 1832, the electric telegraph's instantaneous message transmission speed was shocking. The first commercial television broadcasts in 1939 literally immobilized many people. Far more significantly, many of these communication technologies have spread throughout the cultures into which they were introduced. These technologies are, however, selective information systems. They record and transmit some, but not all, information. The selective nature of these systems means that they can determine what is relevant and significant information, and they function as the mechanisms by which people have come to know the world around them.

In Part III, "Media Cultures," we examine three distinct media cultures or technocultural dramas, and we suggest that these systems can be identified, distinguished, characterized, and understood by their dominant communication these technology. These three media cultures are the *oral culture*, the *literate culture*, the *electronic culture*.

In Chapter 4, "The Oral Culture," we examine how face-to-face oral communication has influenced the organization, social patterns, and values of the majority of the world's nation-states. First, the structural features of orality are described. The institutions and norms fostered and sustained by oral communication are then identified and characterized. We conclude the chapter by considering critics who have analyzed discourse in terms of its oral emphasis and features.

In Chapter 5, "The Literate Culture," we examine how script and the printing press have influenced a host of radical changes in the organiza-

81

tional schemes, social patterns of interaction, and norms of nation-states. First, the structural features of literacy are described. The institutions and norms fostered and sustained by writing and print are then identified and characterized. We conclude the chapter by considering a critic who has analyzed discourse in terms of its literate emphasis and features.

In Chapter 6, "The Electronic Culture," we examine how television and computers are beginning to influence a host of major changes in the organization, patterns of interaction, and values and norms of entire nation-states, groups of people, and individuals. A description of the structural features of electronic communication is the initial part of our examination. Then, the institutions and norms fostered and sustained by television and computers are identified and characterized. We conclude the chapter by considering a critic who has analyzed discourse in terms of telecommunicative and interactive emphases and features.

CHAPTER 4   The Oral Culture

I
n this chapter, we examine social systems that transmit cultural values and norms from one generation to the next predominantly through oral communication. These social systems are "preliterate," and they sustain their sociocultural identity across time through oral and nonverbal face-to-face modes of communication. Certainly, in virtually every society, everyday experiences are conducted through oral and nonverbal face-to-face communication. What makes the social systems examined in this chapter unique is that they sustain their identity from one generation to another not through written, printed, or electronic modes of communication, but solely through oral and nonverbal face-to-face communication. Rather than functioning as exceptional cases, these "oral cultures," as we call them, constitute the norm. Harris (1986) has reported, "Of the thousands of languages spoken at different periods in different parts of the globe, fewer than one in 10 have ever developed an indigenous written form. Of these, the number to have produced a significant body of literature barely exceeds one hundred" (p. 15). In a similar vein, Sanders (1994) has maintained that "of the approximately three thousand languages spoken in the world today, only some seventy-eight have a literature. Of those seventy-eight, a scant five or six enjoy a truly international audience. Literates make up a very small minority of the world's population" (p. 3).

Devoid of writing, print, and electronic modes of communication, human relationships in the oral culture are necessarily governed by verbal communication, nonverbal communication, and the understanding generated by these communication systems. We are fortunate that a host of scholars have investigated several different oral cultures. Many communication researchers and critics have examined ancient Greek oral poetry (Havelock, 1963, 1986; Lentz, 1989; Lord, 1960; Ong, 1977, 1982; Parry, 1987). Others have found evidence of the oral tradition in the ritual literature of India (see, e.g., Edwards & Sienkewicz, 1990), among native American tribes (Ong, 1982; Zumthor, 1990), Russian peasants (Luria, 1976), and contemporary African-American culture (Edwards & Sienkewicz, 1990; Kochman, 1972). In all, orality constitutes the foundation for a specific kind of technocultural drama in a host of diverse societies and in a variety of different historical periods. We are indebted for these contributions, and we readily employ them to provide the generalizations we offer below.

Using the model outlined in Chapter 3, this chapter is divided into four parts. In the first, we identify the structure of the communication technology of the oral culture, and investigate how this communication technology influences what information could and could not be known in such cultures. In the second part, the cognitive or knowledge system of oral cultures is described. In the third part, the social institutions created by and unique to the oral culture are identified. Finally, in the fourth part, an assessment or evaluation of the oral culture is posited.

THE NATURE OF INFORMATION IN AN ORAL CULTURE

Ten modes of communication are used in oral cultures, and determine the kinds of information conveyed among people when they communicate and therefore influence what peoples in oral cultures can and cannot know. These modes will be detailed here.

Sound and Linguistics

The key to unlocking the mysteries of human communication in an oral culture begins with an appreciation for the unique characteristics of sound. Ong (1967, 1982) has contended that understanding the implications of sound can reveal the conditions of human life in the oral culture and the nature of the oral culture itself. Similarly, Havelock (1982) has underscored the importance of sound in Homeric narrative, while Lentz (1989) has isolated a connection between sound and the art of grammar in Hellenic Greece. Likewise, Parry (1987) has identified sound as a distinguishing feature of Homeric poetry.

Three primary assumptions govern our understanding of the role of sound in an oral communication system.

First, sound is evanescent. It vanishes as quickly as it occurs. Accordingly, sound is time-bound, existing only in the here and now. Yet Ong (1967) also argued that this evanescence reveals the role of sound in formatting human experience: "At a given instant I hear not merely what is in front of me or behind me or at either side, but all these things simultaneously, and what is above and below as well. . . . I not only can but must hear all the sounds around me at once. Sound thus situates me in the midst of a world" (p. 129). Sound is thus a human experience more profound than any other sensual system. For example, seeing or touching a locomotive generates an awareness of surface or substance. By contrast, hearing a locomotive creates an awareness of action—that something is happening. Sound invites the perception of activity; it suggests a presence; it signals power. Loudness, quality, pitch, and duration directly influence the intensity of that perception.

Second, sound organizes human perceptions and experiences in a unique and identifiable fashion. Because sound is a here-and-now experience, shared understandings result from the merger between sound and its context (Ong, 1967). Accordingly, in the oral culture, oral communication relies on language that is decidedly, if not inherently, referential. Luria's (1976, pp. 20–47) discussion of perceptual patterns among the Ichkari women, for example, revealed the oral tendency to employ "concrete and object-oriented" language (p. 33). Luria observed that the use of referential language "carries not only meaning but also the fundamental units of consciousness reflecting the external world" (p. 9). In this regard, sound focuses human perception on immediate existence and focuses oral communication on the mood and the tone of the present interaction. In an oral culture, sound encourages and empowers action. Thus, in an oral culture, sound becomes thought in action.

Third, physiologically, sound gains attention and shapes thought by echoing and resonating through the auditory system. In an oral culture, Havelock (1986, pp. 65, 119) has argued, human communication is acoustic. In his view, this oral communication system is primarily an echo system. The echo is an "acoustic law . . . which seems to supply connections as a kind of binding principle which ties bundles of recited situations together" (Havelock, 1982, p. 177). Accordingly, the echo principle primarily functions as a mnemonic device. In an oral culture, with no visual means of recollecting and recording history or events, echoing serves as the gateway to the cultural archives. Ong (1982), for example, contended that "in an oral culture, experience is intellectualized mnemonically" (p. 36). Havelock (1982) has shared this view and maintained that oral composition "eschews the unexpected, prefers the known and familiar, and tends to guide the memories of its audience by anticipation, prediction and responsive echo" (p. 305).

In this sense, echoing encourages two distinct features of oral communication.

First, it engages memory through formulaic repetition. Havelock (1986) has observed, for example, that Milman Parry "discovered tokens of a persistent echo sounded in the recurring formulaic epithets attached to proper names" (p. 51). Indeed, Havelock (1963) speculated that the patterns of the *Iliad* were "built on acoustic principles, which exploit the technique of the echo" (p. 128). Moreover, he suggested that "the familiar formulaic devices of oral technique—the ring form, the repetition with speakers changed, and similar devices . . . all at bottom utilise the principle of the echo" (p. 136).

Second, echoing rhythmically fixes experience in the context of the present. The echo principle is deeply rooted in the fabric of orality. Biological rhythms make the acoustic echo a familiar partner in oral communication. In Havelock's (1982) words, the "echo is something that the ear of the singer and audience is trained to wait for. Its mnemonic usefulness encourages the presence of anticipation" (p. 177). He suggested that all biological pleasures and intellectual pleasures are closely linked to the motor responses of the human body. In fact, Havelock (1986) contended that "acoustic rhythm is a component of the reflexes of the central nervous system, a biological force of prime importance to orality" (p. 72). In other words, the rhythm of a beating heart or the rhythm of breathing predisposes humans to acknowledge and appreciate rhythm in general. Likewise, Gregg's (1984) work on symbolic inducement and knowing is particularly helpful in developing this notion. He contended that the repetition of formulaic devices in oral poetry became important functional parts of "experiencing and meaning" (p. 75). By repeating features of experience, "the rhythmic processes of oral presentation would serve to make such modes of thought more stable, fixing them in rhythmic pattern" (p. 76). In Gregg's view, "all mind–brain activity is rhythmic activity. We perceive data rhythmically, and such data rhythmically pulsate across the pathways of mind–brain." Indeed, he noted that besides enabling speech, "rhythm appears to be central to levels of arousal, to emotional states, to pattern recognition, and indeed to conception and comprehension. Rhythm underlies the process of knowing, the behaviors of social interaction, and can induce symbolic involvement and action" (p. 106). Thus, what might appear to the literate eye as mere echoing or repetition in oral conversation is to the ear a rhythmic intensification of experience. Alter's (1985, p. 29) investigation of the poetry of the Bible reveals that the repetition of phrases, ideas, actions, themes, and images by ancient Hebrew poets amplifies or intensifies them. In other words, rhythmic repetition focuses, heightens, or specifies human experience.

Sound is the foundation for linguistics. People believe and act as if talking affects the meanings conveyed in communication, and there are events where the significance of talk cannot be denied. Yet we have little reliable evidence regarding how valuable the content of verbal communication is in oral cultures. In cultures with print and electronic modes of communication, we know that talking conveys fewer social meanings than nonverbal forms of communication. Mehrabian (1968, pp. 53–55) has maintained that only 7 percent of the total impact of social meanings are conveyed through the content of the verbal mode, while 38 percent of social meanings are conveyed through vocal quality, and 55 percent are conveyed through facial expression. Similarly, Burgoon, Buller, and Woodall (1989) have argued that only 35–40 percent of meaning is conveyed through the verbal mode. In all, such findings at least provide a foundation for considering the impact of nonverbal modes of communication.

Haptics

Haptics is the study of how touch functions as a variable affecting communication. Contemporary researchers suggest that newborns learn about their world through tactile exploration (Maurer & Maurer, 1988). Others, such as Henley (1977), contend that touch can assert authority, hostility, warmth, and sexuality. In an oral communication system, touching could be perceived as a powerful communicative device. Any number of religious rites, for example, emphasize the healing or curative function of the "laying on of hands" (Henley, 1977, p. 96).

Oculesics

Oculesics is the study of how eye contact functions as a variable affecting communication. Eye contact amplifies sound and structures meaning in a number of important ways. For example, Knapp and Hall (1992) have observed,

> We associate various eye movements with a wide range of human expressions: downward glances are associated with modesty; wide eyes may be associated with frankness, wonder, naivete, or terror; raised upper eyelids along with contraction of the orbicularis may mean displeasure; generally immobile facial muscles with a rather constant stare are frequently associated with coldness; eyes rolled upward may be associated with fatigue or a suggestion that another's behavior is a bit weird. (p. 295)

One can well imagine how the oral communicator's effectiveness is enhanced by an appropriate eyebrow flash (Morris, 1977) as the hero of an epic poem is confronted by the unexpected. Similarly, a bored audience's eye behavior could cue the oral poet to pick up the pace of the story.

Aromatics

Aromatics is the study of how smell functions as a variable affecting communication. Smell invites communal integration and participation. Hall (1969), for example, noted that "the external secretions of one organism work directly on the body chemistry of other organisms and serve to help integrate the activities of populations or groups in a variety of ways. Just as the internal secretions integrate the individual, external secretions aid in integrating the group" (p. 34). Thus, newborns and their mothers easily identify one another through olfactory cues (Porter, Cernoch, & Balogh, 1985; Porter, Cernoch, & McLaughlin, 1983; Porter & Moore, 1981). In an oral communication system, these types of olfactory cues greatly enhance social integration. Moreover, aromatics can serve as cultural mnemonics. As Bertelsen (1992) noted, "the smell of meat cooking on a fire could activate collective memories of home and family, or celebratory feasts" (p. 327).

Proxemics

Proxemics is the study of how spatial manipulations function as a variable affecting communication. Sommer (1969) has demonstrated that spatial relationships contribute to social hierarchies and create territorial perceptions that influence interactions. Thus, seating arrangements around a campfire or table might indicate social status. The oral communicator could easily signify social conflict generated by external foes by encroaching on someone's territory. Similarly, spatial manipulations can also signal the degree of intimacy in a relationship. Picture the oral communicator reciting a story about the return of a loved one—closing the distance with the audience to echo the immanent return of the lover.

Kinesics

Kinesics is the study of how body movements function as a variable affecting communication. Kinesics are a natural physiological manifestation of oral acoustic intelligence. Although an infinite variety of gestures and body movements might surface during an oral communicator's performance, we believe the dance would be a most important feature of an oral communication system (Joyce, 1975). Coordinated with linguistics, body movements can constitute an overwhelmingly important factor in understanding what the content of verbal communication means. Smiles, frowns, and hand gestures are only a small portion of the body movements that reinforce, qualify, or deny the content of verbal statements (Scheflen, 1972).

Chronemics

Chronemics is the study of how time functions as a variable affecting communication. The amount of time people are willing to wait, the appropriate time to arrive or leave a social gathering, the length of time taken to plan in advance, and the time of day for certain types of talk to occur are important cultural variables affecting the quality and quantity of communicative exchanges (Knapp & Hall, 1992). During oral conversation, effective pausing, the elapsed time between spoken words, adds emphasis to verbal statements and can also serve as a turn-taking cue. In an oral culture, silence might be interpreted as a delay in response that could indicate a change in mood, subject, or speaker.

Objectics

Objectics is the study of how the use of objects functions as a variable affecting communication. Clothing, jewelry, automobiles, and furniture, for example, all contribute to the potential meanings available in an interaction (Molloy, 1975). Attributes such as wealth, power, and luxury may be assigned to others simply on the basis of costume or possessions (Morris, 1977). Ceremonial dress and artifacts frequently constitute the symbols of authority in an oral culture.

Coloristics

Coloristics is the study of how hue, brightness, and saturation of light function as variables affecting communication. Consumer products often rely on color to identify scent or flavor. In like manner, brighter lighting might be employed to enhance worker productivity in corporate settings (Knapp & Hall, 1992). We anticipate that oral cultures employ similar uses of color, brightness, and lighting when conducting ceremonial rituals such as rites of passage, weddings, and funerals.

Vocalistics

Vocalistics is the study of how the intonation of verbal/linguistic acts functions as a variable affecting communication. The use and quality of the human voice are often used in evaluating personality and in making judgments about others. Moreover, intonation can affect the meaning of verbal statements (McCroskey, 1993, p. 137). Take for instance the statement: Woman without her man, is nothing. A simple intonation of the voice can produce a completely contradictory meaning: Woman, without her, man is nothing. The repetition of formulaic devices, commonplace in an oral culture, relies heavily, although not completely, on vocalistics to generate appropriate situational responses.

These 10 modes of communication do not function as discrete or independent variables during face-to-face exchanges. Rather they interact, and they each function as mutually related variables in an interdependent process. Both in practice and conceptually, these modes are intimately related and mutually define and influence each other in terms of the messages conveyed to others.

In practice, these 10 modes of communication converge in virtually every face-to-face communicative exchange. We are, however, predominantly interested in the relationship between communication and culture and therefore in those special ceremonial events in the oral culture when a storyteller, for example, attempts to convey the values of one generation to the next. The telling of the story, the nature of the storyteller, the environment in which the story is told, the content of the story itself, and the predisposition of the learners constitute the mechanisms for and the effectiveness of the transmission of these values. Based on Fisher's (1984) analysis of the origin of rhetorical forms, we suspect that the narrative form is likely to have been the primary strategy employed to transmit values from one generation to the next. This narrative or storytelling moment would necessarily involve all of the forms of oral communication. The story itself would be verbal, a linguistic presentation, perhaps reinforced by music and verbal tags from older members of the group who already know the story. Additionally, as the story was told, a good storyteller would use intense forms of eye contact, various distance manipulations coordinated to the events unfolding in the story, appropriate and reinforcing gestures and body movements, and dramatic timing. In all, each of the modes of communication common to the oral culture would converge in the telling of the stories that convey and reinforce the values of the society.

In conception, the 10 modes of communication in the oral culture also converge. By way of example, Table 3 indicates how just two of these variables—proxemics and linguistics—are related (see Tubbs & Moss, 1994, p. 109; but see also, for equally important explanations and variations in these factors, Jandt, 1995, p. 77; and Lustig & Koester, 1996, p. 201).

But in a face-to-face interaction, all 10 of these communication forms simultaneously interact and influence the meanings conveyed to others. Given all of the factors involved, a face-to-face interaction potentially can be a tremendously dramatic, intense, complex, and involving experience. When all 10 of these forms interact, people are likely to respond to the impact of the total interaction or to the entire system, rather than to

TABLE 3. Social Distance

Distance	Description of distance	Vocal characteristics	Message content
0–6 in.	Intimate (close phase)	Soft whisper	Top secret
6–18 in.	Intimate (far phase)	Audible whisper	Very confidential
1½–2½ ft.	Personal (close phase)	Soft voice	Personal subject matter
2½–4 ft.	Personal (far phase)	Slightly lowered voice	Personal subject matter
4–7 ft.	Social (close phase)	Full voice with some loudness	Nonpersonal information
7–12 ft.	Social (far phase)	Full voice with some loudness	Public information for others to hear
12–25 ft.	Public (close phase)	Loud voice talking to a group	Public information for others to hear
25 ft. and over	Public (far phase)	Loudest voice	Hailing, departures

any one of these variables. The word *image* has also been used to identify the totality or system that people frequently respond to when engaging in face-to-face communication.

In an oral culture, these images can be understood as the organizing principle or governing structure that determines how data are integrated into socially identifiable patterns. These patterns may reflect and preserve cultural identities, and they may be used to transmit cultural knowledge from one generation to the next. Boulding (1956/1961), an economist, has contended that images locate humans in space, in time, in personal relationships, and in intimations and emotions. Further, he has suggested that "behavior depends on the image" and that "the meaning of a message is the change which it produces in the image" (pp. 6–7).

Images also played a similar role in Greek oral culture. Havelock (1963, pp. 189–190) maintained, for example, that the images conveyed through oral epic poetry visualized experience. In this view, images engage audience and poet alike in an illusion that they are participating in an act being performed. Insofar as visualization was able to capture the attention of an audience, they would be swept along new experiential trails. Thus, the images consciously or unconsciously employed in face-to-face communication encourage social integration by linking sounds and nonverbal behaviors to induce involvement, participation, and shared emotional experience.

Beyond conversation and poetry, music plays an instrumental role in oral communication systems. Rhetorical theory and musical theory share an ancient relationship (Havelock, 1982, p. 14). Gorgias and Aristotle, for example, borrowed terms and principles from music to explain rhetoric and poetry (Medhurst & Benson, 1984, p. 231). In addition, music played a significant role in the educational systems of ancient Greece and other oral cultures. Lentz (1989) has contended that "prior to the end of the fifth century Greek education is largely a matter of choral movement, music, and recitation" (p. 50). Moreover, we believe the perceptual framework necessary to comprehend music closely parallels the perceptual framework employed in an oral culture. Music relies on sound, rhythm, words, and tempo for its communicative efficacy (Bloodworth, 1975); encourages physical responses that make meanings accessible by amplifying emotional appeals (Dunham, 1975; Irvine & Kirkpatrick, 1972); advances and sustains cultural lifestyles and images by encouraging psychological identification (Bloodworth, 1975; Booth, 1976; Christenson & Peterson, 1988; Chenoweth, 1971; Knupp, 1981; LeCoat, 1976); reveals cultural values and attitudes (Bloodworth, 1975; Booth, 1976; Kosokoff & Carmichael, 1970; Mohrmann & Scott, 1976; Stone, 1976); affirms cultural themes and myths (Bloodworth, 1975; Booth, 1976; Campbell, 1975; Gonzalez & Makay, 1987; McGuire, 1984; S. Smith, 1980); and promotes community and communal understandings (Booth, 1976; Knupp, 1981; Lewis, 1976; Roth, 1981; S. Smith, 1980; Thomas, 1974).

| KNOWING IN AN ORAL CULTURE

Thought and expression possess distinctive features in an oral culture. Ong (1982, pp. 37–57) has identified nine features that distinguish the oral culture from other ways of understanding.

1. Thought and expression in the oral culture are *additive rather than subordinative*. Oral communication systems do not rely on formal grammatical principles and analytical linguistic structures to determine meaning. Instead, existential contexts encourage a pragmatic oral structure that ignores internal relations between ideas. The result is an additive structure that strings ideas together rather than subordinating them to one another. For example, in Book 21 of the *Iliad* (Homer, 1990) after Hera implores Hephaestus to end his attack on Xanthus:

She ceased
and the god of fire quenched his grim inhuman blaze
and back in its channel ran the river's glistening tides,
And now with the strength of Xanthus beaten down
the two called off their battle. (p. 532)

This string of ideas, although contributing to the oral narrative, disregards a subordinative relationship between those ideas. Thus, oral communication systems are more responsive to the improvisational requirements of the communicator than to the syntactic demands of the narrative (see, e.g., Edwards & Sienkewicz, 1990, pp. 39–48).

2. Thought and expression in the oral culture are *aggregative rather than analytic*. Oral intellection requires formulaic grouping of events to assist memory and recollection. Images are conjured by clustering formulaic elements like clichés, epithets, and themes. In the oral memory, "majestic mountains" are remembered as such, not as mountains to which majestic qualities can be attached. Breaking up thoughts into analytic elements decreases the likelihood that those thoughts could be remembered.

3. Thought and expression in the oral culture are *redundant or copious*. The physical conditions that empower the echo principle demand a redundant form of thought and expression. Since sound is evanescent, there is no external linear continuity to aid the memory—there is no returning to a previous thought. Consequently, repetition assures intellectual continuity in oral communication systems. Thus, creative repetition encourages a stylistic copiousness in oral expression.

4. Thought and expression in the oral culture are *conservative or traditionalist*. If knowledge is to be retained in an oral culture, it must be continually used and repeated. Thus, past experiences, accumulated wisdom, and those who conserve and sustain the "social encyclopedia" are highly valued (Havelock, 1963). As a result, oral communication systems favor traditional knowledge and discourage free mental play. Well-worn phrases and old sayings such as "a rolling stone gathers no moss," or "children should be seen and not heard," or "let he who is without sin cast the first stone" are regularly employed to characterize everyday experience. Oral communicators might vary the language used to express this sort of traditional wisdom, but they would of necessity remain true to the inherent meanings of such messages.

5. Thought and expression in the oral culture are *close to the human life-world*. Lived experience serves as the foundation for oral-based thought and expression. Thus, human agents and their actions provide the conceptual ground for knowledge in an oral culture. For example, history would be recorded as a genealogy, a listing of events according to

human relations. Instructions for building a boat or strategies for defending the city from attack would not appear as a simple list of technical procedures, but could be presented as remembered human activity. Accordingly, information of this sort would be more easily retained by presenting it as an account of the adventures of a great seafarer or warrior hero.

6. Thought and expression in the oral culture are *agonistically toned*. Because knowledge in an oral culture is tied to human experiences, oral expression adopts the spirit of a dramatic conflict between human agents. Existence in the experiential world of orality forces recognition of the immanent struggle with nature and with others. By casting knowledge as lived experience, oral communication systems engender the agonistic tone inherent in such human existence. Thus, the oral communicator relates the account of the seafarer in contest with the sea and the wind, or the warrior in contest with the swinging sword and cunning enemy. Accordingly, the agonistic tone permeates the verbal performance and is institutionalized in the lifestyle of the oral culture.

7. Thought and expression in the oral culture are *empathetic and participatory rather than objectively distanced*. The mode of interaction in oral cultures is decidedly social and participatory. Since oral communication relies on sound and nonverbal behaviors to construct shared images of experience, face-to-face interactions between immediate sources and immediate receivers are essential if communication is to occur. From this perspective, feedback is immediate, continuous, and constant. The mimetic nature of oral communication precludes the possibility of a subjective sensibility. Instead, oral knowing and learning depend on an empathic communal interaction in which the knower and the known are interdependent and inseparable. Thus, active identification with a story and its characters engages both oral communicator and audience in an experiential event that transcends biological isolation and social estrangement. As a result, oral communication systems establish the community as an epistemic system that maintains social bonds through individual participation.

8. Thought and expression in the oral culture are *homeostatic*. Oral cultures are profoundly dependent on memory to sustain cultural traditions, knowledge, and lifestyles. Information and meanings committed to memory define social relationships and are directly tied to existential events. Thus, the continuity of social relationships and, correspondingly, the continuity of an oral culture are constituted by a delicate balance between relevant information and the ability to remember it. However, social relationships could be expected to change over time. Retaining information no longer necessary would burden the memory and disrupt cultural equilibrium and continuity. Accordingly, forgetting information

that is no longer useful would be as important a process as remembering. Indeed, discarding irrelevant information contributes to cultural homeostasis by freeing the mind to focus on present events.

9. Thought and expression in the oral culture are *situational rather than abstract*. Meaning in an oral communication system is decidedly context-specific. Since the lived experience provides the conceptual framework for orality, communication is tied to everyday reality, and messages are constructed and received in terms of present events. Accordingly, oral communication relies on an operational framework of interpretation rather than an abstract one. Thus, members of an oral culture would characterize objects in terms of their practical application rather than classify them as components of a specific categorical class. For example, bowls, cups, forks, and knives would be defined within the context of their use in eating a meal— you put your food in a bowl, you drink from a cup, you cut your food with a knife, you pick up your food with a fork—rather than as members of categorical classes such as dishes and utensils (see, e.g., Luria, 1976).

Clearly, the basic units for analysis of oral communication systems, the 10 forms of verbal and nonverbal communication as well as the integrated sense of the image that they convey, establish unique interactional patterns for orally based thought and expression. Facilitated through conversation, poetry, and music, these components taken together characterize the communication system in an oral lifestyle. In terms of the traditional rhetorical canons, delivery and memory ultimately become the controlling and regulatory variables, perhaps even dogmatic ones, in their understood importance, in the preservation of the cultural identity in an oral culture. In addition, these production components identify delivery and memory as the rhetorical canons most characteristic of oral communication systems (see, e.g., Chesebro, 1989; Ong, 1982).

THE SOCIOLOGY OF THE ORAL CULTURE: SOCIAL INSTITUTIONS AND INDIVIDUAL EXPERIENCES

Oral cultures generate a host of constructs that dominate human communication. Functioning as a form of cultural lag, oral culture constructs may even be carried over into literate and electronic cultures, prescribing oral culture ideas, images, interaction patterns, and values that can remain viable, credible, and persuasive in literate and electronic cultures. Insofar as face-to-face communication remains a powerful factor in the repertoire of human communication techniques, we expect that these oral culture constructs, to one degree or another, will remain with us.

In oral cultures, the significance and power of the community is elevated, and it becomes a dominant social institution of the oral culture. The immediate, ever-present community is essential for face-to-face communication to occur, and this community identity constitutes the most important way of understanding the self. In oral communities, the notion of community specifically relies on *mythos* or the shared attitudes and beliefs of the community. In an epic poem, for example, the poem's hero personalizes, concretizes, and idealizes the community, but, far more importantly, the hero functions as a model of the culture's values and beliefs. As the hero encounters choices affecting the welfare, if not the existence, of the community, these choices can assume a moral flavor that, by the end of the story, constitutes and reflects the perfected ideals of the community.

In this context, rather than shifting the orientation or leading a community, a hero mimes the community and becomes a singular, condensed symbol representing it. Accordingly, in an oral culture, a hero extends and reinforces the values of a community. *Mimesis*, the rhetorical figure of social integration through representation, characterizes this symbolic role of the hero in an oral culture (Chesebro, 1989).

Similarly, for the vast majority of community members, the community itself creates an active, personal identity. The self is understood in terms of the responses of others in the community. Without the community, there can be no definition of self. The self achieves meaning within the context of the community's definitions, functions, and values. In all, the oral community establishes the grounds for morality, gives priority to value systems, establishes a scheme for assigning roles to its members, and creates a sense of being that all members of the community share.

Beyond elevating community as a primary social institution, the oral culture establishes a second "social institution" that is more precisely understood as a sociological norm. Particularly, the participatory nature of the oral community makes concrete and sensory experience the most powerful mode of evidence for a claim. Indeed, the importance attributed to the concrete may be the origin of and explain why people consider an "objective" view to be possible and important. Luria (1976) has provided evidence for the power of using concrete references when communicating in an oral culture. In this regard, his oral and literate culture comparison is revealing:

> Situational thinking was the controlling factor with illiterate peasants from remote areas who farmed the land alone and had never spent any time in a large city. On the other hand, our second group of subjects— people who either had taken short courses or had become involved in the communal work of the newly organized collective farms—had

reached a certain transitional stage. They were able to employ categorical classification as an alternative to practical grouping. (p. 68)

In addition to relying on concrete references, everyday communication is the most trustworthy and persuasive mode of communication in
the oral culture. Certainly, everyday communication reflects the lived
experiences of oral culture members; during such interactions, meanings
are derived from immediate contexts and therefore are part of what is
immediately perceived and "understood." Less reliable forms of communication would be derived from, in William James's (1869/1950) typography, "orders of reality" or "sub-worlds" that include the world of science
or of physical things as the learned conceive them, the world of ideal
relations or abstract truths, the prejudices common to a race, the various
worlds of individual opinion, and the worlds of sheer madness and vagary
(pp. 292–293). By contrast, everyday communication constructs a "reality" in which particular intentions rather than intentionality are treated
as an epistemic issue. A single reality is assumed to exist independent of
perception; imperative actions are believed to be required rather than
dialective exchanges; interactions contain the full scope of all verbal and
nonverbal stimuli; members assume that there is a correspondence among
their meanings; and it is assumed that relationships will continue to exist
in terms of geography, time, and social relations (Chesebro & Klenk, 1981,
pp. 327–328). In all, for members of the oral culture, their daily talk with
members of their own community constitute the most trustworthy and
persuasive mode of communication.

Additionally, in the oral culture, synchronistic feedback constitutes
a dominant mode of interaction, and this type of feedback functions as
the primary mode used to understand. Asynchronistic feedback controls
alternative systems of communication. In these systems, an uninterrupted
presentation of materials is employed. Understanding is accordingly derived by paying careful attention to how logical developments, sequential
orderings, and categories are linked. By contrast, in the oral culture, the
give-and-take of continuous and mutual exchange of verbal and nonverbal stimuli generates meaning. In this sense, as a mode of communication,
music can only be appreciated and "understood" if the listener "partici-
pates" in the musical experience on a physiological, immediate, continuous, and constant basis.

Moreover, in the oral culture, messages are constructed and transmitted
in the present, with the past and the future understood within this present
orientation. In face-to-face interaction, as noted at the outset of this chapter,
verbal and nonverbal symbols are "evanescent." They vanish as quickly as
they are transmitted. Understandings are linked to the moment when they
were first generated and produced. The past exists, then, only insofar as it

can be re-created in the present with contemporary understandings. Similarly, conceptions of the future are tied to verbal and nonverbal communication symbols in the present. In all, as a symbolic orientation, a timeless quality exists in the oral culture, the urgency of change is tied to present understandings, and change becomes an extension of the present.

Furthermore, in the oral culture, the process of knowledge is linked to a specific interaction in which a knower (the source of knowledge), the known (what is known), and the learner are a product of a face-to-face relationship. Fisher (1984) has suggested that the oldest and most powerful mode of communication is the narrative form. In an oral culture, the narrative predominantly occurs when a storyteller (the source of knowledge), the story (what is known), and the listener (the learner) interact in an immediate, face-to-face, synchronistic interaction.

Finally, in the oral culture, mysticism and creativity are linked to referential language usage. Discussions of purpose, ends, and objectives are necessarily linked to concrete and sensory experience. Abstract goals are immediately and operationally defined in terms of concrete and sensory experiences. In this regard, an abstract entity such as "nature," that possesses no phenomenal characteristics, can only be understood in terms of its physical manifestations, such as hurricanes, lightning, droughts, and floods.

In all of these ways, oral cultures generate powerful social institutions and individual experiences. Indeed, these social institutions establish the norms, patterns, and strategies of interaction for specific individual communication acts.

EVALUATING COMMUNICATION IN AN ORAL CULTURE

The modes of communication creating and characterizing the oral culture can exert powerful influences on the social organization and sense of self. The verbal and nonverbal forms as well as the totalistic impact of these forms as a coherent system or image is psychologically and physiologically involving, intense, and participatory. Nevertheless, communication systems relying on oral culture techniques warrant, as much as any other mode of communication, critical reactions that can emphasize both the advantages and limits of this type of communication, particularly in terms of the value of these techniques in a society that seeks to enhance the quality of symbol using.

The advantages and disadvantages of the oral culture are revealed in the social institutions created and sustained by its modes of communication.

For us, the oral culture possesses a host of intriguing possibilities. It creates a sense of community and gives the individual a place in the community; it provides a language that possesses clear standards for its referential usage; it promises a world of known and objective clarity with precise meanings. Further, it provides a "reality" in which lived experience and immediate contexts function as a foundation for human communication. A sense of involvement and direct experience is created through its use of synchronistic feedback, which allows knowledge to be linked to a known source and specific learning experience. We must admit that at times we seek to recapture this cultural experience.

However, the oral culture possesses characteristics that we would avoid in practice. It is a cultural system in which individuals can become "flat." The inner motives of specific individuals are deemphasized, and individual motives are tied to their role in the community. Community objectives outweigh individual objectives. The individual is conditioned to accept these community definitions, particularly through the use of rhetorical strategies such as mimesis.

Additionally, the hierarchies that permeate the oral culture disturb us. In the stories we have from oral cultures, hierarchies are readily employed and seldom questioned. The elders of the community function as leaders by virtue of their accumulated memory. Yet these hierarchies also have been destructive. In ancient Greece, for example, some 95 percent of the population were slaves, and women were denied a public forum for speaking (Kennedy, 1963). We are thus concerned that the hierarchical nature of oral cultures possesses a propensity toward oligarchies rather than toward representative and participatory forms of democracy (Bertelsen, 1992).

In all, as critics of the oral culture we are attracted by its promises, its sense of the immediate and the concrete. Although we would not generalize at this time, we must admit that we are hesitant to commit to such a system predominantly because it appears that a few benefit in these cultures at the expense of a larger majority. But at this point, we now turn to the reactions of other critics who have analyzed oral cultures.

Because of preservation problems, oral culture critiques are not readily generated or available. Because oral conversation relies on sound as a basic unit of analysis, an actual critical example would of necessity be as evanescent as the object of its critique. Nonetheless, insofar as Homeric poetry is often taken as a written record of oral conversation, we believe a critique of this form provides an important illustration of how the communication techniques of the oral culture can be assessed. Accordingly, we have selected two analyses to illustrate how critiques of the oral culture might be handled.

Oral Poetry

The first is an analysis of oral poetry as reflected in the *Iliad*. Specifically, we have chosen Knox's (1990, pp. 3–64) introduction to the *Iliad*, contained in the translation by Fagles, to illustrate the nature of media criticism of oral culture communication. Knox's introduction often wavers between identifying the *Iliad* as oral conversation and as written literature. Indeed, Knox (p. 7) has been persistent in his belief that the *Iliad*, as we presently have it, could only be the product of oral and literate communication systems. Accordingly, although much of his introduction speaks directly to the *Iliad* as an example of oral conversation, he also examined its literate dimensions. Nevertheless, we believe that his explicit and implicit references to orality and oral communication systems provide sufficient information to illustrate how an oral communication system can be analyzed as a technocultural drama. Additionally, in terms of this discussion, we have found it convenient to organize our remarks in terms of the model of media criticism proposed at the end of Chapter 3.

Structural Analysis

Explaining the technology involved, Knox (1990) identified production components consistent with those of an oral communication system. He implicitly employed sound as a basic unit for analysis. For example, he explored the notion that the *Iliad*, like most epic poetry, was probably sung in performance (p. 7). Indeed, a great deal of his analysis focuses on the rhythmic nature of hexameter verse (pp. 5, 15) and its role in the selection of formulaic phrases. Knox noted that meter in the *Iliad* is based on pronunciation time rather than syllabic stress (p. 12). Accordingly, he suggested that ornamental epithets and the formulaic nature of whole themes were dictated by the poem's meter. Moreover, Knox contended that the oral communicator's improvisational skills relied on a set of metrical alternatives that allowed the poet to select formulaic phrases that fit the poem's meter rather than its narrative (pp. 15–17). As a result, the *Iliad* exhibits the verbal repetition, or recurring formulaic epithet vital to the stimulation of oral memory. This formulaic echoing thus marks the *Iliad* as a memorable form, one that transmits and preserves information by engaging the audience's communal imagination (pp. 8–9, 12, 24). Further, Knox's recognition of the importance of formulaic repetition characterized the aggregative and redundant nature of oral thought and expression.

Because of Knox's tendency to view the *Iliad* as literature, he often overlooked the role of nonverbal behaviors in the production of oral messages. However, there are some oblique references to nonverbal behaviors that appear from time to time. For example, Knox acknowledged that the *Iliad*'s characters engage in face-to-face interactions (pp. 21, 55). In addition, he observed that oral communicators adapted to their audience, no two performances being alike (pp. 19–20). Audience adaptation and face-to-face interactions implicitly suggest that nonverbal behaviors would play an important role in constructing messages in an oral communication system. Insofar as the epic poem is part of the social encyclopedia of an oral culture, heroes engaging in face-to-face interactions are modeling a behavior common to the culture. Likewise, audience adaptation depends in large part on the oral communicator's ability to recognize and respond to an audience's nonverbal reactions.

Knox also identified nonverbal echoes of the epic poem in Greek art vases of the period (p. 7) and in dance (p. 62). Thus, nonverbal cues and artifacts constitute a significant element in the oral communicator's repertoire and in the oral culture's inherited materials.

Images comprise another unit of analysis in Knox's introduction to the *Iliad*. He contended that Homer's epic poem commemorates the heroic image of the modern world (p. 29). Moreover, Knox suggested that this image encourages the audience to celebrate heroic values such as honor, courage, and devotion as they share in the heroic myth (p. 27). Implicitly, Knox also found that the *Iliad* invites the audience to share visions of social order and hierarchy (p. 48), thus re-creating for them the distribution and structure of power in ancient Greece. Moreover, Knox made his readers increasingly aware of the *Iliad*'s capacity to model appropriate attitudes (p. 32) like rage (p. 38), sorrow (p. 60), and anger (p. 53). In a more general sense, Knox displayed the poem's ability to convey traditional information that contributes to social unity and integration. His discussion of funerals and sporting contests reflected the ritual enactment of cultural customs essential for social bonding (p. 57).

Cognitive Analysis

The cognitive dimension of Knox's (1990) introduction is also consistent with what we know about an oral communication system's role in shaping human apprehension, thought, and understanding. Knox found, for example, that the *Iliad*'s primary appeal is centered in its close relationship to the lived experience. He maintained that destructive violence is made familiar by "drawing for illustration on the peaceful, ordinary activities of everyday life" (p. 62). Moreover, Knox demonstrated that human agents and their actions show the audience what motivates the *Iliad*'s characters rather than describing that motivation in narrative fashion (p. 47). Thus, the *Iliad* provides its psychological insights through the lived experience that we have identified as the foundation for oral-based thought and expression.

In addition, because the *Iliad* relies on lived experiences to engage the audience, it exhibits an agonistic tone consistent with the experiential world's pervasive struggles between humans and with nature. Essentially, the *Iliad* is a poem about war: war between city-states (p. 24), war between emotions (p. 56), war between lifestyles (p. 64). The audience is swept into the poem and shares the experiences of its characters as if they were its own. Thus, human thought adopts the agonistic tone of the epic and human existence is institutionalized in the context of the present situation.

Knox also provided evidence that suggests oral communication systems rely on a process of forgetting to maintain homeostasis. For example, in his explanation, the mixture of dialects in epic language displays the active forgetting process that maintains cultural equilibrium by discarding irrelevant information. As he illustrated, the sound represented by *w* "disappeared from the Greek alphabet early on, as the consonant ceased to be sounded in the spoken language" (p. 13).

Recognizing the potential range of individual and contradictory meanings that can be extracted from the *Iliad*, Knox's introduction clearly illustrated how competing cultural systems and technologies influence understandings of the *Iliad* as well as the range of possible reactions and responses to it. His discussion of literate and oral interpretations of the *Iliad* indicated that it is possible to read the poem as a literal history of ancient Greece, or embrace it as a part of an oral culture's shared vision of the agonistic struggle for humane existence in a barbaric world (pp. 3–22, 62).

He also suggested that for the audience in an oral culture, the *Iliad* could encourage a broad range of identifications and interpretations. If the lived experiences of the poem's hero, Achilles, personify appropriate social conduct, then the audience would be invited to experience a gamut of emotions and perspectives that indulge in the heroic myth and, conversely, deny the myth. Audiences might revel in Achilles' anger and fury as he slaughters those who killed his friend, yet if the moral of the epic is recognized, they would also come to appreciate the ultimate futility of violence and death. Thus, when Achilles comes to a new way of thinking, the audience might do so as well (p. 60).

Sociological Analysis

Knox (1990) suggested that the *Iliad* sustains a number of social processes indicative of oral societies. For example, he noted that the *Iliad* creates a vision of civilized life (pp. 31–32). This vision casts the *polis* or city as an important element of Greek life (p. 30). The communal aspects of life in the Greek city-states emphasize the importance of the community; it is cast as surviving at the expense of the individual (p. 34). Indeed, the city-states in the *Iliad* regularly sacrifice the lives of their finest citizens so that the city might endure. Moreover, direct empathic participation represents the only viable link between what is known and the knower. Achilles cannot appreciate the pain and suffering felt by his enemies' survivors until he loses his own friend. Likewise, the audience cannot appreciate the *Iliad* unless they are consumed by its hypnotic effect (p. 12).

In Knox's view, the *Iliad* also sustained social institutions such as the family. He contended that the Greek concept of family is embodied by the gods (p. 41). More importantly, the gods personify mysterious life forces. When events are beyond human comprehension and control, those events can be attributed to the whim of the gods (pp. 42–43). In this way, audiences are not required to logically explain experience; they simply live those experiences. The phenomenon is appropriately reflected in small children who do not understand why they are forbidden to touch the family's collectibles; it is enough that their parents (gods) have commanded them not to. Thus, the family or the gods supply a form of consciousness that does not exist in the preliterate mind (Jaynes, 1976).

Evaluative Analysis

Knox's (1990) evaluative analysis of the *Iliad* suggested that through empathic identification with the poem's heroes, the audience is reminded of what it means to be human—to feel sorrow and regret, to comprehend the consequences of rage, to be ostracized from the community and become a part of it again (pp. 51–58). Indeed, Knox argued that the *Iliad* is not merely the celebration of brutality and death inherent in an agonistically toned poem about war and warriors. Instead, he contended that contemporary audiences of the *Iliad* should throw off traditional perspectives of it as a history of the violence in ancient Greek life, and embrace it as a recognition of human values and of an abiding respect for civilized human life (pp. 61–64).

Rock and Roll Music

In addition to Knox's analysis, we have selected Grossberg's (1986) essay "Is There Rock after Punk?" as an example of a critical analysis of music conceived and designed to reflect an oral culture. We find this essay a compelling choice for two reasons. First, "Is There Rock after Punk?" reveals the close relationship between music and oral communication systems in general. Second, Grossberg's essay provides a slightly different perspective on the relationship between technology and culture. Although technology is generally cast as influencing culture, Grossberg's essay occasionally asks its readers to consider the converse: how cultural systems influence technological forces.

Structural Analysis

Grossberg (1986) implicitly acknowledged sound as a structural element of music. He has contended that the content of rock and roll lyrics do not necessarily signify meanings. Instead, he has argued that rock and roll relies on "sheer volume and repetitive rhythms" to "produce a real material pleasure for its fans" (p. 52). Thus, rock music rhythmically fixes or contextualizes experience in much the same way as oral conversation. Musical sound, regardless of its genre or its mode of production, repeats and amplifies rhythms that focus human experience. In this way, music functions mnemonically to echo experiential and emotional conditions in both audience and song.

In oral conversation, rhythm often encourages corresponding non-verbal behaviors. Grossberg, in passing, has suggested that nonverbal behaviors, such as dance (p. 56), comprise a structural component of music. Moreover, he devoted significant attention to the importance of the body in this essay. His discussion of the body and the use of sexuality in rock and roll music clearly indicate the influence of nonverbal behaviors in constructing images of youth culture.

Grossberg employed the image as his primary structural unit of analysis. He has maintained that images of youth and the body (or pleasure) provide "the unity and continuity of rock and roll culture" (p. 56). Indeed, from Grossberg's perspective, the celebration of youth centered in the pleasures of the body simultaneously defines rock music and inscribes the affective context in which it is heard. In all, in his analysis, communication systems affect cultural systems, and cultural systems affect communication systems.

Cognitive Analysis

As a way of understanding, Grossberg (1986) has cast rock and roll music as a "mode of functioning" (p. 50) that empowers its fans by organizing the fragmented images of postmodern living into coherent structures of desires and relations (p. 55). Moreover, he has argued that the effects of rock and roll should not be examined simply as a definition of audience, but as part of the broader cultural context as well, and as a part of the affective experience of its fans (p. 54). Rock and roll music thus becomes an attitudinal response to the structures of everyday life shared by music consumers. Accordingly, Grossberg found that rock and roll music creates an "affective alliance" or shared emotional and attitudinal perspective between audience, music, and culture (pp. 56–57). In this way, rock and roll evokes psychological identifications and transactions through the lived experience.

Moreover, Grossberg's conception of lived experience recognizes the possibility that additional layers of complex understanding can be involved. He has maintained that rock music encourages or relies on an oppositional politics to isolate a specific audience from the broader sociopolitical context. In this way, rock and roll encourages a shared experiential or emotional perspective by operating as a cognitive strategy that illuminates "competing and contradictory ideological practices" of experience within a broader social context that shapes and is shaped by experience (pp. 54–55). Thus, for example, rock and roll might be seen as youthful resistance to the discipline of family and schools. But in the broader sociopolitical context, rock and roll becomes a communicative form that empowers the struggle of youth against the constraint of oppressive social institutions and identities (p. 58). In this view, rock and roll transcends everyday life and exposes traditional structures of power. Given the added complexity Grossberg's analysis has provided of everyday experience, an oppositional politics emerges that is clearly reminiscent of the agonistic tone inherent in oral communication systems.

For Grossberg, punk rock represented a distinct musical form that encourages a postmodern reading of music and culture. He has noted, for example, that different people may use and interpret rock and roll music differently (p. 52), often in unintended ways. The essence of Grossberg's argument is that rock and roll music organizes everyday experiences into structures that challenge existing power relationships and promote personal desires. Accordingly, punk rock is cast as a postmodern response to existing social institutions and identities, including rock and roll music itself. Punk rock challenges rock and roll's assumed representation of social and emotional experience, and criticizes rock and roll for being out of touch with postmodern youth. At the same time, punk rock responds to postmodernism by rearticulating the images of youth and body (pp. 58–62) as they relate to postmodern society. Thus, punk rock encourages broader interpretations of social conditions and is largely constructed by those same conditions. In short, punk rock recognizes that "all social life is constructed" (pp. 60–61) and is itself constructed from the strands of contemporary existence.

Sociological Analysis

In terms of its impact on societal structures and processes, Grossberg's (1986) essay again reaffirmed the close relationship between music and oral conversation. When its functional impact is recognized, rock and roll music relies on empathetic audience participation for its effectiveness. Thus, when conceptualized as a lived experience, the affective alliance establishes a transactional relationship between what is known and the knower. Audiences do not appreciate rock and roll music through a signifying narrative framework. Instead, rock and roll engages its audiences at the level of emotional and physiological association. This form of direct participation is consistent with the communal participation characteristic of oral cultures. As a result, rock and roll music acts as a bonding mechanism that articulates and sustains its audience as a community.

Grossberg has particularly observed that a number of specific social institutions are also directly influenced by rock and roll. In fact, he argued that punk rock alters or challenges several of these institutions. For example, punk rock is, in many ways, a response to capitalism. Punk's challenge to capitalism manifested itself in a shift from large, dominating record companies to a host of smaller companies, where the production of inexpensive single records encouraged new groups and new releases (p. 58). Yet punk did not limit capitalism. Instead, recording companies, as members of a capitalist social institution, merely co-opted the unique qualities of punk. Thus, punk, as a response to the postmodern vision for capitalism, ultimately sustained the institution, albeit in a slightly modified version.

Other social institutions are also challenged, but also ultimately sustained, by punk. The family and school as well as images of youth and body have all undergone a metamorphosis when viewed as the social ingredients of punk musical constructions. Indeed, punk music is constructed from and reflects new social conceptualizations and organizations. As Peterson (1987) has suggested, punk challenges the despair of the fragmented postmodern world by organizing affective responses to changing cultural conditions. Thus, the family and schools, once viewed as protective environments where children could learn how to become adults, are reconceptualized as oppressive institutions that victimize children through physical and emotional abuse. Accordingly, Grossberg contended that rock and roll's inherent antidomesticity is self-threatening (p. 67). Although punk rock might challenge existing institutions such as the family by exhibiting the very qualities it opposes, punk brings its own viability into question. In these circumstances, punk is defined by the affective standards it opposes. As a result, punk implicates rock and roll as a constraining structure of everyday life.

Evaluative Analysis

Grossberg's (1986) evaluation of rock and roll centers on whether it can still be viewed as affective empowerment and as oppositional (p. 69). He concluded by suggesting that cultural critics must ask if rock and roll is "working affectively for particular audience fractions and how is it positioned within hegemonic politics?" (p. 69). Accordingly, he contended that "rock and roll is a set of strategies for struggle" (p. 69) that should operate at three levels: everyday life, affective, and hegemonic (p. 70). In sum, Grossberg viewed rock and roll as the empowerment of "the changing structures of our affective existence" (p. 70). Rock and roll music attempts to demystify the constraining influences of power and relational structures by creating affective alliances that bridge the gap between everyday life and the ideological contradictions of contemporary culture.

| CONCLUSION

In this chapter, we examined communication in an oral culture. In such cultures, 10 forms of communication can be distinguished, but these forms operate simultaneously, often producing overall images that define and characterize human communication in the oral culture. This type of communication affects what is defined as relevant information and what people ultimately identify as knowledge. This mode of communication has exerted an overwhelming impact on people. Several social institutions can be traced to societies that predominantly use oral modes of communication, and these institutions appear to possess amazing staying power. We have concluded this chapter by summarizing the work of critics who have focused on the essential nature of orality and music to render assessments and judgments. The examples in this chapter suggest that oral communication, as a discrete technocultural drama, is a complex system amenable to interrelated structural, cognitive, sociological, and evaluative analysis.

CHAPTER 5

The Literate Culture

In *The Name of the Rose*, Eco (1983) sought to capture the drama of life in a literate culture. In his fictional monastery, solitary monks spend countless hours in quiet contemplation trying to discover the meanings embodied in the text of ancient manuscripts, often adding their own thoughts to a growing library collection. Indeed, at the moment one is reading, the rest of the outside world fades away, and the reader is alone, solitary, and separated from others within the community. Beyond the social relationship created by reading, Eco also captured how people understand, reason, and argue within a literate culture. When reading, the only available "data" are words, in prescribed sequences. These words are abstractions, categories, genres, or groupings referring to specifics. The prescribed sequence among these abstractions has no relationship to the specifics being examined. Thus, reflecting the arbitrary and conventional nature of literacy, for example, the word *dog* is an abstraction referring to all barking animals, some of which are four-legged, but the word also refers to those barking animals who, because of accidents, are only three-legged. In the case of words such as *honor* or *love*, the categorical nature of a word is more obvious, and it becomes far more significant how the reader reacts to the word and the kind of example he or she associates with the word. Indeed, for a single word to be meaningful, the reader must use the words to mentally construct an example of the referent or experience the word is designed to represent or constitute.

The characteristics attributed to literacy differ sharply from those of the oral tradition. In the oral culture, for example, both speaker and listener provide simultaneous and constant feedback that enhances the communication process. However, both writer and reader need not be present for literate communication to take place. Additionally, meanings in an oral culture are negotiated through recognition of, identification with, and exchange of commonly held assumptions and expectations about language, culture, and reality. The corresponding relationship between speaker and listener reflects the intimacy and immediacy of the moment in an everyday environment. By contrast, meanings in a literate culture are, at best, indirectly negotiated between writer or reader through the words of the text. It is unlikely that the writer and reader will ever directly interact. Without situational context and the immediate feedback available in face-to-face interaction, meaning becomes subject to individual interpretation. Understandings in a literate culture can and do develop independent of an interaction involving the writer and reader's presence, character, and intention. Thus, literacy or the use of literate communication technologies contributes to a technocultural drama that differs significantly from social systems generated by orality.

Using the model outlined in Chapter 3, this chapter is divided into four parts. First, we identify the structure of the communication technology of the literate culture and identify how this communication technology influences the kind of information that is and is not known in such cultures. Second, the cognitive or knowledge system of literate cultures is described. Third, the social institutions created by and unique to the literate culture are identified. Finally, an assessment or evaluation of the literate culture is posited.

THE NATURE OF INFORMATION IN A LITERATE CULTURE

Literate communication systems are characterized by a reliance on visual rather than acoustic perception. The use of physical notation systems, such as pictographs, ideographs, syllabaries, or phonetic alphabets, generate visual representations of language. In general, visual representation systems have played a dramatic role in the development of writing and print (see, e.g., Coulmas, 1989; Illich & Sanders, 1988; Logan, 1986).

The demands of visual perception largely determine the influence of literate communication systems. In addition, the applications to which literate communication systems are put contribute to significant changes in the communication process. Quite simply, visual representation systems, often commonly understood as abstractions of human sound, introduce new elements into human communication. Such elements include the notions of visual perception, linear and sequential processing, the text as context, and abstract thought. Consequently, the structural features that distinguish the written mode from the oral mode constitute an appropriate starting point for any analysis of literacy as a technocultural drama.

Visual Perception

The appropriate role of the visual in human communication has been an enduring issue. Some 2,500 years ago, Plato used the term "sight-seers" to refer to the average Greek citizen. The term was decidedly pejorative, as Plato believed "sight-seers" were unable to engage in abstract thinking that would lead to more exact knowledge (in Havelock, 1963, pp. 240–251). From Plato's perspective, *doxa*, or opinion derived through the memorization of oral poetry, constituted a confused mental state that was decidedly inferior to the more exact knowledge available through abstract identification. Thus, "sight-seers," or those bound to an oral tradition, grounded cognition in the "poeticized experience of narrative events," while philosophers, or those adopting a literate tradition, grounded cognition "in the context of sense experience of physical objects" (Havelock, 1963, p. 249).

Responding to the visual emphasis of the contemporary literate culture, a host of scholars have argued that literate technologies are accompanied by a shift in perceptual modes, from the broader use of all the senses in an oral culture to a particular focus on visual mechanisms in literate cultures (see, e.g., Havelock, 1963, 1980; Innis, 1951; Logan, 1986; McLuhan, 1962; Ong, 1967). Specifically, McLuhan (1962, p. 39) and Derrida (1967/1976, p. 3) have maintained that the phonetic alphabet reduces speech to a visual code. As such, literacy dissociates the alphabet from physical experiences (see also Logan, 1986; Ong, 1982).

This shift from the oral–aural to the visual took thousands of years to accomplish. Ong (1982) has succinctly summarized the shift from oral–aural perceptual modes to visual ones: "All script represents words as in some way things, quiescent objects, immobile marks for assimilation by vision. . . . The alphabet, though it probably derives from pictograms, has lost all connections with things as things. It represents sound itself as a thing, transforming the evanescent world of sound to the quiescent, quasi-permanent world of space" (p. 91).

Once the eye replaces the ear as the primary receptor, significant changes occur in the ways humans process events around them. Quite simply, sight imposes different constraints on cognitive processes than does the spoken word.

More specifically, three significant perceptual differences can be discerned between sight and sound.

First, sight privileges space and spatial relationships. Ong (1982) has argued that the alphabet reduces "sound directly to spatial equivalents" (p. 91). As a result, the alphabet significantly alters the way in which humans interpret their world and the events around them. As Ong (1982) has explained, the alphabet implies "that a word is a thing, not an event, that it is present all at once, and that it can be cut up into little pieces, which can even be written forwards and pronounced backwards; 'p-a-r-t' can be pronounced 'trap' " (p. 91). In all, the phonetic alphabet's tendency to focus on sight rather than sound fragmented the Greeks' perceptions of the world around them. Instead of the multisensual experience of actuality conveyed through an oral tradition, the Greeks began to envision the world in a fragmented manner, where events could be and were separated from their lived context and examined independently (see, e.g., Logan, 1986, pp. 121–122).

Second, the intensification of visual perception also focuses attention on exterior phenomena. Oral traditions centered the individual in an acoustic field where space "is not spread out in front of us as a field of vision but diffused around us" (Ong, 1967, p. 163). In the multisensual world of oral–aural perceptions, inner rhythmic harmony constitutes human experience. Visual experiences, however, deemphasize oral ways of knowing. Coulmas (1989, p. 47) has argued that alphabetic writing's accuracy in representing the spoken word increases in proportion to the particular alphabet's segmentation of the spoken language. This segmenting of experience is traceable to the use of visual perceptual modes. Moreover, Gregg (1984, pp. 98–102) has convincingly shown that humans have an innate capacity to visually segment the flow of external events. More importantly, Gregg has indicated that the phonetic alphabet provides the resources for socializing visual segmentation. In his view, the physiological conditions to support a visual communication system are naturally occurring. It is not surprising, therefore, that humans would develop a representational scheme that collaborated with and celebrated its physiological foundation. Thus, the natural human ability to visually segment experience by relying on spatial relationships instead of auditory–tactile ones finds expression in a written form that functions in a similar manner. In both cases, exterior phenomena play an increasingly important role in meanings developed through literate communication technologies.

Third, by privileging space over sound and by focusing on exteriors, sight also encourages a profound sense of order and control. Oral communication systems encourage direct experiencing of the human life-world. Within the rhythmic harmony of human existence, events are communicated through shared physiological and psychological states. For example, members of an oral culture might reasonably share the fear and panic a forest fire generates. Their collective understanding of such an event, induced through story or song, would be largely determined by the storyteller's or singer's ability to evoke specific psychological and physiological conditions. This direct experiencing of reality encourages the perception that events happen over which humans have little control. Quite simply, life is experienced as it happens.

Literate communication systems, by contrast, heighten awareness of external realities. In order to communicate effectively about external phenomena, those phenomena must be named. The process of naming randomly occurring exterior phenomena, to make sense of the exterior world highlighted through vision, identifies what is named and conveys a degree of power to the enumerator (see, e.g., Ong, 1982, p. 33). Beliefs of this sort are common in oral cultures but are often dismissed as superstitions. However, it seems reasonable to assume that the practice might carry over to literate cultures as well (see, e.g., Kaufer & Carley, 1993, pp. 375–380). Havelock's (1963) discussion of Plato's Forms provides a particularly revealing example of how the process of naming imposes a sense of order and control over the physical environment. Havelock has argued that Plato chose to symbolize moral abstractions by referring to them as Forms (pp. 262–264). He has suggested that Plato believed, in a generic sense, that the Forms "enjoy a kind of independent existence: they are permanent shapes imposed upon the flux of action" (p. 263). In this way, Plato emphasized the importance of understanding the structure and logic, or order, of the external world. At the same time, he revealed an understanding of visual tendencies to symbolically fix and hold reality. Such understanding implicitly suggests that inherent in the act of naming is the concurrent sense of order and control over natural phenomenon (see, e.g., Ong, 1967, p. 136).

In sum, literate communication systems supply the visual resources that symbolically restructure what the physical environment is understood to be. The process of making symbolic sense of the world through the use of literate communication technologies imposes regularity on uneven events because the concepts of order and control inhere in the nature of symbol usage itself. As Gregg (1984) has argued, "symbolic activity induces by imposing order" (p. 134).

The intensification of visual perception encouraged by literate communication systems privileges spatial relationships, focuses attention on exterior phenomenon, and encourages a sense of order and control in an otherwise haphazard experiential world. As a result, literate communication systems significantly influence how information is processed and understood.

Linear and Sequential Processing

An increased reliance on visual perception encourages a sequential construction of the world and its events. Where sound relies on the simultaneity of experience to organize thoughts and perceptions in the context of the present, visual perception is nonsimultaneous. As Ong (1967) has noted, "As a human being, I see only what is ahead, not what I know is behind. To view the world around me, I must turn my eyes, taking in one section after another, establishing a sequence" (p. 128). Explaining his position, Ong has noted that vision "is a dissecting sense. Or, to put it another way, one can say that it is sequential. It presents one thing after another. Even though each part of the landscape surrounding me on all sides is contemporaneous with every other part, sight splits it up temporally, gives it a one-piece-after-another quality. . . . The actuality around me accessible to sight, although it is all simultaneously on hand, can be caught by vision only in a succession of 'fixes' " (p. 129).

From this perspective, concepts expressed in the written mode are to be understood in a linear or sequential format, following the order of the words in a sentence or on a page. Literate communication systems thus facilitate a linear organization of thought (Ong 1980, p. 199). For example, when walking into a restaurant the ambience of the dining room is taken in holistically; as a collaboration of aromas, lighting, decorations, and sounds. Writing a sentence to describe that experience transforms a moment of simultaneous actuality into a temporal sequence. With a beginning, middle, and end, the form of the sentence itself offers a sequential or linear organization of the information it presents. Because concepts understood visually are cognitively organized as sequences, literate communication systems simulate the actuality of present time. Accordingly, concepts expressed in the written mode can transcend the immediate space and time in which they are created.

Text as Context

The earliest forms of Greek literacy were reproductions of oral poetry, but the meanings derived from the written text and the spoken word differ significantly in the way they are constituted.

Oral meanings are arrived at through a face-to-face negotiation and are intimately connected to the actual setting. As a result, the concept of *text*, as a communicative event isolated from its sociohistorical context, is unknown in oral cultures. In contrast, literate meanings are primarily derived through a negotiation between writer or reader and a text. In this situation, meanings can develop independent of their setting and without the presence of others. Moreover, if literate meanings are consistently developed without benefit of context, it seems likely that idiosyncratic interpretations of texts would be commonplace, and we would expect the emergence of a movement, such as postmodernism, that assumes that every symbolic act inherently conveys multiple and contradictory meanings. From a literacy perspective, such an assumption can be, for some, extremely difficult to accept because the literate culture assumes that meaning and understanding are shared. Accordingly, the relationship between text and context becomes a focal point for understanding within a literacy orientation.

In this regard, two perspectives have emerged regarding the relationship between text and context.

First, written texts, unlike oral messages, are often encountered apart from the sociohistorical context in which they were created. As a result, a number of scholars have argued that literate communication systems produce messages or texts that are context-free. Ong (1982), for example, has suggested that "written words are isolated from the fuller context in which spoken words come into being" (p. 101). In addition, Scribner and Cole (1981) have demonstrated that schooling that is based on information in texts is associated with context-independent forms of reasoning. Finally, we have Coulmas's (1989, pp. 144–150) discussions of North Semitic script and Hebrew orthography that indicate that negotiating the meaning of early writing often involved the notion that the text was itself the context for meaning and understanding.

Second, written messages omit substantial information that would be used to aid comprehension in oral communication settings. Facial expressions, gestures, and subtle variations in pitch, tone, and volume are not easily represented in literate communication systems. As a result, scholars who share Ong's focus on the text as a weak replication of the spoken word commonly hold that in literate cultures the text exists independent of a sociohistorical context. Olson (1977, 1991b), for example, has argued that many written texts are autonomous. In texts such as these, meanings are dependent on the wording of the text and have little to do with the cultural context in which the text was created. The text itself supplies the context for interpretation. In this way, the text can be thought of as both text and context. As Sanders (1991) has suggested, from an ontological perspective, texts themselves constitute a reality.

By contrast, texts can be understood as cultural artifacts. Street (1984) has argued that every text embodies a set of linguistic conventions and ideological assumptions that are consistent with and reflect the cultural applications of any given literate communication system. Literacy requires some knowledge of the conventions of writing and reading as well as some broader awareness of literate traditions—the uses and applications of texts in social situations and institutions. These conventions and assumptions signal a general awareness of extratextual context. In this regard, Ong (1982) seems to have implicitly recognized this relationship, for he has noted that some texts are able to supply extratextual information that aids in the vocal replication of the text. From this perspective, extratextual contexts play an important role in literate communication systems.

Abstract Thought and Language

In ancient Greece, literate communication systems created and promoted abstract thought (see, e.g., Havelock, 1963; Logan, 1986; Ong, 1982). This position has been argued in a variety of ways. Havelock (1963) and Lentz (1989) have suggested that abstract thought was not possible in Greek oral culture. In their view, only through the use of literate technologies could abstract thought develop. For example, Lentz (1989) has argued that "writing constantly reinforced the awareness of symbolic, abstract relationships. Writing was the symbol for the spoken word, which was, in turn, the symbol for thought" (p. 4). As Lentz has explained, "People grew aware of categories as they reread and puzzled over relationships between written words, words writing removed from the intonation and situations that make them vocally concrete and contextually specific. Humans developed awareness of abstract relationships between categories as they reread and revised versions of texts" (p. 4). As a result, abstract reasoning was considered more trustworthy than observation and was fully available through competence in a literate communication technology.

Moreover, other researchers have often articulated a relationship between literacy and abstract thought by drawing a causal connection between literate communication systems and abstract thinking. Specifically, literate technology itself is commonly identified as the source of abstract thought. Logan (1986, p. 21), for example, has maintained that because the phonetic alphabet has the fewest number of signs for spoken language, it is the most abstract of writing systems (see also Coulmas, 1989, p. 268). Logan (1986) has further contended, "All spoken words are abstractions of the things they represent. The written word is a further abstraction of the spoken word, and phonetic letters give it an even greater abstraction than ideographs or pictographs" (p. 104). He has concluded, "The use of the alphabet thus involves a double level of abstraction over a spoken word because the transcription of a spoken word takes place in two steps. A spoken word is first broken up into semantically meaningless phonemes or sounds, and the sounds are then represented by semantically meaningless signs, the letters of the alphabet" (p. 104).

Insofar as literate communication systems are imprecise representations of spoken language, they are inherently abstractions. However, the natural abstraction of the written from the spoken cannot be directly translated as a causative factor in a shift in cognitive processes. To do so oversimplifies the emergence of more explicit abstract thinking in literate societies. Not surprisingly, such a deterministic view of an autonomous technology inducing new cognitive processes has encountered some resistance.

A growing body of research suggests that abstract thinking is not media-specific. On the contrary, studies have shown that some forms of abstract thinking are possible in any language system. Street (1984), for example, contends that "to speak a language at all is to employ abstraction and logic" (p. 26; see also Street, 1984, p. 36). Moreover, Denny (1991) has shown that preliterates do engage in rational or decontextualized thought. The use of literate technologies, then, cannot be the "source" of abstract thought. Indeed, Street (1984) has claimed that "all languages have the potential to make abstract, relatively neutral statements, if called upon to do so" (p. 80).

At the same time, psychologists Scribner and Cole (1981) have investigated the causal connection between literate communication systems and abstract thought. Their findings indicate that nonliterates as well as literates were able to successfully classify and abstract concepts. They have contended that schooling in the use of literate technologies, rather than the technology itself, is largely responsible for making abstract thinking explicit. They have identified three aspects of abstract thought. The first is the "ability to single out a particular attribute (say, size) shared by a set of objects in order to subsume them into distinct, mutually exclusive classes (literally, to *ab-stract* one feature from many as a basis for classification)" (p. 118). The second is the "ability to shift flexibly from one attribute to another (say, from size to shape) as a basis for constructing classes (to be relatively unconstrained by preexisting organization and to envisage other possibilities, sometimes equated with context-free thinking)" (p. 118). And, the third is the "ability to give a verbal label to a class (name it) and to explain the basis for class membership (sometimes designated as verbal-logical thinking)" (p. 118).

Each of these aspects of abstract thought is present in both literate and preliterate cultures. However, these aspects were more explicit in proportion to the amount of schooling available. Thus, the relationship between literate communication technologies and abstract thought has been softened. Instead of claiming a causative role for literacy in the development of abstract thought, scholars are beginning to recognize that schooling in literate technologies heightens or intensifies the process of abstraction. In sum, being schooled in the use of literate technologies invites familiarity with the process of abstract thought. Accordingly, abstraction becomes a more obvious cultural practice. As Street (1984) has suggested, because of their use in specific cultural contexts, "some languages pay less explicit attention to abstract formulations" (p. 80).

Regardless of the conceptual relationship drawn between literate communication technologies and abstract thought, two important matters have gone unchallenged. First, those cultures that rely on literacy as a mode of mediating human communication exhibit more explicit involvement in abstract thinking than do oral cultures. In fact, Denny

(1991) has suggested that abstract thought or decontextualization is a distinctive feature of Western society. Although it cannot be said with certainty that literacy promotes abstract thought, it is important to acknowledge the more explicit nature of abstract thinking in the presence of literate communication systems. Second, literate communication systems are by their very nature abstractions. Written languages represent imprecise replications of spoken languages. In this sense, literate communication technologies do engage abstraction as a cognitive process.

Visual perception, linear and sequential processing, the text as context, and abstract thought and language constitute essential structural features of literate communication systems. Historically, although the written mode is considered an extension of orality, it goes beyond orality and emphasizes different elements in human communication. Orality and the written mode are related but distinct communication media. The differences between the two affect a number of interactional patterns, cultural practices, and social institutions.

| KNOWING IN A LITERATE CULTURE

Thought and expression possess distinctive features in a literate culture. Because they frequently function in different ways in social systems, we have found it useful to distinguish writing and print as modes of communication.

Writing

In contrast to the cognitive techniques employed in the oral culture, Coulmas (1989, pp. 11–14) has identified six alternative techniques for understanding that stem from the nature of writing itself: mnemonic, distancing, reifying, social control, interactional, and aesthetic. These functions directly affect cultural conventions, social relationships, and social institutions. By examining each of these functions, the differences in interactional patterns between literate cultures and oral cultures are clarified, and the broader implications of the written mode emerge.

First, writing serves a mnemonic function. While orality relies on rhythmic narrative to store information and aid recall, the written mode substitutes a fixed text (see, e.g., Yates, 1966). Oral memory was primarily concerned with the present, collecting and recollecting information relevant to the current situation or experience (Havelock, 1982). The specific spoken language of orality had a fleeting existence. Once spoken, words could not be recovered. Cultural cohesion and knowledge was maintained by employing rhythmic repetitions of long-standing myths as a primary pedagogical device. Accordingly, departures from cultural myths as explanations for current events were unlikely.

Writing, on the other hand, gives language a degree of permanence. The written word remains exactly as written, unchanged by time or events. The permanent nature of written memory dramatically increased both the type and the amount of information that could be recalled (see, e.g., Coulmas, 1989). Moreover, the spatial dimension of the written mode strongly appealed to the emerging focus on visual perception. Thus, written documents slowly gained public acceptance and ultimately were privileged over the spoken word (see, e.g., Clanchy, 1979; Goody & Watt, 1963; Stock, 1983). The written record, correspondingly, provided the resources for accumulating knowledge and encouraged the study of history (see, e.g., Coulmas, 1989; Goody & Watt, 1963).

Second, the written mode distances writer and reader spatially and temporally. Oral communication relies on face-to-face interactions between speaker and listener. In contrast, written communication may occur even though writer and reader are separated by geographic distance or live in different historical eras. As a result, the social relationship between speaker and listener in oral cultures is altered in major ways by the introduction of writing. The social relationship of orality reflects the immediacy of the moment of the exchange, while the social relationship between writer and reader can occur without the presence of one or the other. Similarly, the written mode diminishes or delays feedback. Readers separated by time and distance from writers cannot engage in the simultaneous and continuous feedback that characterizes orality. Literate meanings can exist independent of the presence, intention, and context of the writer and reader. Thus, writing and reading are isolated experiences through which a sense of individuality emerges. Indeed, "family," "community," and "society," all become concepts that can be understood independent of the individual (see, e.g., Eisenstein, 1979; Huizinga, 1954; Manchester, 1992).

Third, the written mode reifies or objectifies language. As Coulmas (1989) has noted, "The spoken word is ephemeral and spontaneous in its very essence. In writing, on the other hand, words become stable and tangible" (p. 12). The spatial constraints of visual perception, inherent in the written mode, invigorate language itself with a physical existence. Because writing encourages readers to envision language as an object, concepts expressed in the written mode can be examined discretely and as individual entities apart from their context.

As a knowledge system, the written mode allows for a sense of objectification in which the human being is cast as and understood as independent of the environment and its controls. Importantly, the context-free nature of the written mode invites deduction, abstraction, rationality, and the critical analysis of language. Olson (1977, 1986, 1988; Olson & Torrance, 1987) has long held that the objectification of language through writing encourages interpretation as a way of knowing and understanding the meaning of a text (as opposed to rhythmic harmonious experiencing in an oral culture). Moreover, he has suggested that interpretation as a function of writing is in large part responsible for the emergence of fields such as hermeneutics and science (see also Stock, 1983).

Fourth, the permanence of the written mode has the potential to regulate social conduct (Goody, 1986). Olson (1991b) has argued that the definition of literacy as "an individual's ability to read and write" should be broadened to include "the general competence required to participate in a literate tradition" (p. 252). Four essential conditions are necessary to support this literacy hypothesis: (1) There must be some method to fix or preserve and accumulate texts; (2) there must be institutions for which the use of texts is relevant; (3) there must be institutions that induct learners into institutions that use texts; and, (4) an oral metalanguage must evolve that allows authors and readers to refer to texts (Olson, 1991b, pp. 253–254).

Competence with this oral metalanguage, and with the use of writing and reading, privileges and empowers a literate class. In fact, most researchers agree that the written mode contributed to a new class system (see, e.g., Clanchy, 1979; Havelock, 1982, 1986; Illich & Sanders, 1988; Lentz, 1989; Logan, 1986; Ong, 1967, 1982; Stock, 1983). For example, orality requires a storyteller and listener, but both must participate to create the experience. Stories can be retold by anyone. However, the written mode requires special skills that are not easily acquired and that can be withheld from others, encouraging a "literate class" that can function as a "privileged class."

In addition, the written mode altered the regulatory mechanisms of orality. The shared "ethos" of the oral community was replaced by codified law. As Coulmas (1989) has claimed, "Social control is exercised through writing because the 'letter-craft' has always been carried by privileged elites who could refer to written documents as seemingly objective standards of human conduct" (p. 13). In the oral culture, a community was bound by time, space, and the regularity of contact. But because codified laws could be exported and applied in the absence of authority, the written mode supported geographically unrelated communities. Moreover, the written mode provides a collaborative mechanism for preserving control and order in a large, complex, and multilayered urban environment (see, e.g., Coulmas, 1989; Goody, 1986; Scribner & Cole, 1981). The written mode thus provides social cohesion and control.

Fifth, the written mode functions interactionally by increasing the potential for human action. Writing requires formal and universal education or training with artificial technologies. The technologies of orality are "natural," or built into the individual (speaking requires the human voice and listening, the human ear), whereas the written mode requires "artificial technologies" external to the body of the writer and reader (e.g., the book, the pen, and paper).

Familiarity with writing as a tool of literacy thus "makes possible novel kinds of coordinated action" (Coulmas, 1989, p. 14). For example, using the alphabet also involves the ability to classify information (Logan, 1986). In this way, the written mode can be employed to create and to organize unrelated facts, while orality is constrained and governed by ideational unity and coherence. Dictionaries, encyclopedias, self-help books, and wills, for instance, all indirectly mediate human interactions.

Sixth, the written mode possesses its own unique aesthetic standards. Clanchy (1979) and Stock (1983) have maintained that copying written manuscripts was an art. Decorative marginalia, calligraphy, and special parchment were carefully employed to create manuscripts that appeal "both to the intellect and the sense of visual beauty" (Coulmas, 1989, p. 14). Thus, the written mode encouraged new forms of understanding based on visual aesthetics. Light/dark metaphors, genres such as novel and drama, and concepts such as speculation, reflection, and illumination emerged as literate knowledge systems (see, e.g., Manchester, 1992). As literate communication technologies became privileged over orality, style and arrangement became the critical rhetorical canons (Chesebro, 1989).

Clearly, the structural features of the written mode privilege ways of knowing that are different from those in oral cultures. Visual perception, sequential and linear processing, text as context, and abstract thought and language contribute to and emphasize specific interactional patterns characteristic of literate thought and expression. Employing literate technologies as a communicative art mediates human experience and, at the same time, shapes and constrains that experience. The written mode functioned in this manner for centuries. With the development of the printing press, the structural features of literacy were transformed and the effects of literacy were intensified and expanded.

Print

Exploring a library provides a profound sense of the difference between oral, written, and print-based communication systems. Stacks and stacks of books, neatly ordered, sit silently on the shelves. This arrangement suggests the uniformity and regularity commonly associated with print-based communication systems. However, until books are read, they say nothing. The message conveyed by books does not exist until the solitary scholar selects and reads them. Acting in relative isolation, the reader is required to supply the appropriate context for interpreting the text. The richness of the oral tradition is gone from the pages of the book. No matter how artful the author, the book is devoid of the drama and shared ecstasy of orality. For example, the story of Genesis in the Bible is hard pressed to convey the physiological and psychological conditions of the Creation. Readers do not experience these conditions directly. Instead, a listing of events simply records and recalls a history of creation.

In ancient writing-based cultures, manuscripts were read aloud and functioned as mnemonic devices to assist the oral presentation of events (see, e.g., Clanchy, 1979; Havelock, 1982; Lentz, 1989). The communicative act still retained much of the drama of the oral experience and the promise of direct personal interaction. Attendance at professional conventions attests to the continuing popularity of writing as an artistic inducement for oral interaction. But print-based cultures are unable to maintain the slightest pretense that they embody the holistic understandings of orality. Only insofar as the reader imagines a conversation with a book's author does the print mode replicate or preserve the oral experience.

Multiple forms of print exist and have been traced historically from the use of cylinder seals by the Sumerians and the use of carved reversed wooden blocks by the Buddhists in sixth-century China (see, e.g., Coulmas, 1989; Eisenstein, 1979; Logan, 1986; Unwin & Unwin, 1987). The use of movable type with wooden blocks was embraced in Europe and, when influenced by alphabetic literacy, generated a movable type font with one alphabet letter per font. The printing press utilized this form of movable type font to mass-produce manuscripts and books at low cost (see, e.g., Logan, 1986). Although some controversy exists, the invention of the printing press in the 1440s is commonly attributed to Johannes Gutenberg (see, e.g., Eisenstein, 1979).

An entire cluster of social, economic, historical, political, religious, educational, cultural, scientific, and military factors affect the development, acceptance, and use of a new technology. Logan (1986), for example, has illustrated some of the complex social transformations that encouraged the development of printing in his assessment of the demand for books in the Middle Ages. He has specifically suggested that there were "three distinct periods for the production of manuscripts" (p. 179): (1) The need for the monasteries to preserve classical texts and organize the scriptoria to carry out copying tasks from A.D. 550–1200; (2) the need for the universities to stock books required for particular courses from A.D. 1200–1400; and (3) the needs of commercial publishing houses, starting in 1400, to distribute religious books, satisfying the emerging literati and the university market.

The printing press emphasizes the general perceptual and cognitive processes of the written mode. However, the printing press is not just a special subform of the written mode. The printed and written modes differ as basic types, modes, or media of communication. Printing represents a significant shift in literacy style. Logan (1986, pp. 176–226) has identified four features of the print mode that distinguish it from the written mode. In our view, these differences profoundly affect how we understand and know.

First, print standardizes a text by virtue of the mass production of fonts, the regularity employed in setting letters and lines, and the uniformity in page formats. In contrast, the written manuscript was a work of art, by virtue of the individual handwritten script employed to generate the manuscript. Ong (1982, p. 122), for example, has suggested that white space on a page plays an important role in manuscript and print cultures. He has contended that in the written mode, the control of space is ornamental. But in the print mode, the control of space conveys regularity. The spacing between words, between different marks of punctuation, and at the beginning of new paragraphs directs the reader's attention in a consistent and regular pattern (see, e.g., Bolter, 1991). Indeed, manuscripts were hardly read in the strict linear fashion that books are. Manuscript "reading" was a mnemonic exercise rarely consistent with a verbatim translation of the text (see, e.g., Stock, 1983).

The regularity of page formats combined with the improved legibility of type fonts to produce texts with an undiversified appearance (see, e.g., Havelock, 1982). Written manuscripts depended on the artistic talents of the individual copyist. Insofar as the copyist could represent words clearly, the manuscript would function as an appropriate mnemonic for oral reading. Often, content or wording could differ dramatically from one copy of the same manuscript to another. But the printed page, uniform in format, offers an unchanging, precise representation of the text. Reading a printed book thus standardizes language in a way that orality and manuscripts could not.

Second, print makes texts accessible and permanent. The time, skill, and energy required to produce a written manuscript meant that written manuscripts were scarce, limited in number, and inaccessible to most of the population (see, e.g., Clanchy, 1979; Eisenstein, 1979; Stock, 1983). The speed of typographic duplication made automatic reproduction possible (see, e.g., Clanchy, 1979; Havelock, 1982). The result was a dizzying increase in the number of copies of a text, thereby making the text accessible to a broader segment of the population (Unwin & Unwin, 1987). Moreover, typographic printing was relatively inexpensive. The speed of production and the sheer quantity of books that could be printed from one typesetting was passed along in the form of reduced prices for books. Low prices and an adequate supply made books an accessible source of communication in literate societies. Further, the printing press made it possible to use vernacular languages, increasing readership and promoting literacy (Logan, 1986). Manuscript writing, commonly performed by religious or academic scribes, employed languages such as Latin to restrict readership. By reproducing material through the phonetic alphabet, print could meet the demands of a growing readership on their own linguistic grounds.

In addition, the rigid visual fixing of the printed word assures the endurance of the text (see, e.g., Havelock, 1986; Ong, 1982). By fixing language in a medium where it can be repeated consistently, regularly, and uniformly, print renders language permanent (see, e.g., Clanchy, 1979; Havelock, 1986). Manuscripts, although fixed representations of language, were commonly presented orally. The presentation of the manuscript was subject to any number of transformations. For example, the reader's marginalia might become part of the presentation and might, as a result, become part of future copies of the same manuscript. Printed texts were immune to this type of alteration. The permanent nature of printed texts made them more desirable as evidence in court proceedings. Since printed texts were immune to forgery, their credibility was less questionable than manuscripts or oral testimony (see, e.g., Clanchy, 1979; Eisenstein, 1979; Stock, 1983).

Third, print dramatically increases the literate emphasis on visual perception and spatial relationships. Because of the ways print standardizes a text and lends an air of permanence to language, visual communication with a stronger emphasis on spatial relationships was privileged over oral communication and face-to-face relationships. The irregularity of written texts made them difficult to read. In addition, their structural ties to orality invited an oral reading. The lack of punctuation, inconsistent spacing between words, and inconsistent spelling of words demanded that manuscripts must be sounded out if they were to be understood (see, e.g., Clanchy, 1979; Febvre & Martin, 1990; Havelock, 1982; Huizinga, 1954; Lentz, 1989; Stock, 1983). Reading a manuscript aloud commonly required the presence of an audience.

Books, on the other hand, encourage silent reading. By accurately and immutably fixing data in a spatial mode, printing permitted reading without the presence of others. It was no longer necessary to have others present to mediate the communicative act (see, e.g., Olson, 1991a). In this sense, print shifts communication from an oral to a silent reading activity (see, e.g., Clanchy, 1979; Ong, 1982; Stock, 1983). The isolated reader, deliberately apart from others, reading silently, and negotiating the meanings of a text independently, became the norm for reading.

Fourth, print, in collaboration with alphabetization, redefined notions of education from understanding to processing large quantities of information. As Logan (1986) has noted, the printing press changed the patterns of scholarship, shifting from an oral culture's concern with a general understanding of an event to the collection and organization of large "amounts of data in a systematic and orderly fashion" (p. 196). The printing press made it possible to accumulate knowledge beyond the individual memory. Volumes of information were gathered and stored. Referencing systems such as concordances and indexes appeared to facilitate retrieval of printed material. Dictionaries and encyclopedias were produced in vast quantities (see, e.g., Eisenstein, 1979). Early dictionaries gathered commonplace materials, such as epithets, that could be stitched together in oral presentation or manuscript writing, much like the oral poet's "stitching" together of commonplaces to move the audience. These dictionaries often offered the oralist a phonetic pronunciation of words (see, e.g., Ong, 1977). As printing became more and more a part of life, the attention to phonetic accuracy continued. But to it was added a concern for word definition. Since readers engaged a text in isolation, interest in what to say and how to say it began to compete with concerns for what the words meant.

Print also freed literates from the laborious task of reproducing manuscripts and enabled them to devote more time to developing new ideas. Tycho Brahe's vast contributions to astronomy are often cited as an example of the free mental play that the printing press makes possible (see, e.g., Eisenstein, 1979; Logan, 1986). Because books were permanent, they were also considered more accurate repositories of information. Misspellings and typographical errors were relatively easy to identify in the repetitive regularity of the printed word. Irregularities in the text could be edited and corrected, and, as a result of increased accuracy, printed books became a principal focus of instruction in literate societies (see, e.g., Clanchy, 1979; Eisenstein, 1979; Stock, 1983).

These structural features contribute to a number of social constructions, movements, and institutions that are not developed in oral or manuscript cultures. In fact, a host of social institutions and philosophical movements commonly identified with the modern period are regularly attributed to print or reinforced by print (see, e.g., Logan, 1986). Some of those print-dependent social constructions and their influence on interactional patterns will be considered here.

THE SOCIOLOGY OF THE LITERATE CULTURE: SOCIAL INSTITUTIONS AND INDIVIDUAL EXPERIENCES

Eisenstein (1979), a historian, has argued that print "altered methods of data collection, storage and retrieval systems and communications networks used by learned communities throughout Europe" (p. xvi). In her view, these alterations resulted from the influence of print technology on a multitude of concurrent events. Eisenstein does not suggest that print or the printing press is solely responsible for the changing interactional patterns of print-based Europe from A.D. 1400–1700. Rather, she has maintained that the printing press orchestrated and brought together a host of other social developments that led to or encouraged social transformations. The artistic knowledge necessary to develop the printing press, for example, clearly demonstrates Eisenstein's thesis. Knowledge of screws, levers, paper, inks, chemistry, and metal casting contributed to the development of print (see, e.g., Logan, 1986). Much of this information existed prior to the development of the printing press. Indeed, artists and craftsmen practiced these skills long before Gutenberg's press was invented. The interactional patterns and social transformations of print-based culture were not caused by print. Instead, social developments were influenced and encouraged by a merging of events and arts that culminated in the invention of the printing press. As we consider the interactional patterns of print-based culture, it is this perspective we wish to encourage. Print orchestrates a different cultural harmony.

Several social transformations that alter interactional patterns are commonly attributed to print. For example, mass education and mass literacy are facilitated through the printed book. Manuscript writing assured the existence of a literate public long before the invention of the printing press, but greater access and use of vernacular languages invited a much broader readership (see, e.g., Chartier, 1987; Clanchy, 1979). As Logan (1986) has noted, print provided a quick and easy means to diffuse ideas. Since reading was an isolated individual act, readers could be self-taught. For example, Abraham Lincoln is regarded as a self-educated man. This form of education and corresponding literacy make ideas available to a much broader, spatially and temporally diverse audience.

In addition, print contributes to a cultural environment that invites visual thinking and encourages visual arts. Because print intensifies the visual bias of the alphabet, visual thinking emerges as a counterpart to oral thought. The contributions of print were organization, alphabetization, systemization, and standardization (Logan, 1986), which manifested themselves through an interest in the visual arts, such as architecture, geometry, painting, and sculpture (see, e.g., Eisenstein, 1979; Havelock, 1982; Manchester, 1992; Stock, 1983). Moreover, the influence of visual thinking and visual arts invited new intellectual movements and endeavors.

For instance, the structural features of print encouraged and increased the trend toward abstraction, uniformity, classification, and analysis. Not surprisingly, the rise of science, or the sciences, is often linked to the repeatability, preservation, and accuracy of the printed page (see, e.g., Eisenstein, 1979, 1980; Havelock, 1982). The visual propensity for encountering events in a discrete, linear fashion fostered a critical disbelief among members of literate societies. Cultural myths that could not fully account for natural phenomena were increasingly discounted as explanations or rationales for social collectivity. In their place emerged an ever-increasing reliance on natural observation and repeatable experiment. This empiricism fixed knowledge just as immutably as type fonts and ink fixed language on a page. As a result, early Western scientists were able to construct a broad, unifying and unified "scientific" vision of the world by collecting and contrasting a variety of separate, published accounts of natural events. Each individual scientist's contribution thus became, more or less, the appropriate rule of law for the phenomena being considered (Eisenstein, 1979).

The endurance of the Renaissance was sustained in large part by the printed book. A humanistic revival of classical influences was tied to the need to meet readers' increasing demands for new material. The source of this new material was often the written manuscript. In fact, most of the printing press's early efforts were reproductions of manuscripts. The publication of these documents and their distribution to a vast new audience rekindled an interest in classical learning (Eisenstein, 1979, pp. 163–302).

Further, the religious debates of the Reformation in 16th-century England were often conducted through print (Manchester, 1992, pp. 127–219). With a growing number of readers, it was no longer necessary to rely on the established clergy to interpret religious doctrine. In fact, clerical interpretations regularly maintained the status quo in social relationships. Without traditional clerical interpretations, people began to think more independently and formed their own opinions on spiritual and moral matters. As a result, a distinct division between civil authority and faith emerged during the Reformation (Febvre & Martin, 1990; Logan, 1986, p. 219).

Consequently, the printing press also played an instrumental role in the rise of nationalism. Reformation opposition to the Italian papacy, for example, grounded itself in vernacular languages that encouraged ethnic identification (see, e.g., Logan, 1986). The printing press played a key role in disseminating vernacular interpretations of religious doctrine and in promoting national loyalties to an emerging readership. As a result, churches were organized according to national interests and loyalties. This situation was, according to Logan (1986), partly responsible for "national languages and cultures through which the aspirations of nationalism could be expressed and in which a national consciousness could emerge" (p. 223).

At the same time, individual achievement was accorded heightened recognition. Authoring and publishing an essay or a book was, and is, commonly considered a claim to intellectual property rights (see, e.g., Kaufer & Carley, 1993). Moreover, to promote sales of their books, early printers and authors regularly promoted and publicized themselves (see, e.g., Eisenstein, 1979). As a result, the cultural climate that sustained literacy simultaneously sustained a growing individualism (see, e.g., Illich & Sanders, 1988).

Clearly, writing and print encourage technocultural dramas that differ significantly from those engendered through and by orality. Epistemically, literacy separates experience from reasoning about it. The shift from orality to literacy in Western societies, according to Stock (1983), spurred a knowledge shift from "the raw data of sense or the platonized ideal of pure knowledge" to "the forms of mediation between them" (p. 531). As a result, literate understanding emerged "from the accumulation of reiterated and reinterpreted experience" (p. 531). Onto-logically, the permanence of writing and print supersedes the transience of oral custom. Indeed, literacy promoted the "belief that within the ontology of the printed word lay an intimate reflection of reality, which the study of grammar, syntax, and hermeneutics could reveal" (p. 530). Criticism that encounters human events in a technocultural drama shaped through literate media should, of necessity, embrace these ontic and epistemic distinctions.

EVALUATING COMMUNICATION IN A LITERATE CULTURE

Writing and print revolutionalized human communication, and they have permanently altered what and how people know as well as the social organization of human societies. Without writing and print, it is unclear that mass education, the visual arts, the Scientific Revolution, the Ren-aissance, the Reformation, and the concept of *individuality* could have occurred. These benefits, especially if we are predominantly concerned about enhancing the quality of symbol using, are profoundly unchallenge-able. Indeed, the more pressing task is to determine all of the ways these new institutions have positively affected the quality of life for human beings.

At the same time, literacy remains an ability of a minority throughout the world. As we reported in Chapter 1, the vast majority of world cultures rely predominantly on orality, and they transmit their identity and values from one generation to the next primarily through oral communication. Literacy remains a communication technology of relatively few cultures throughout the world. This fact introduces a host of important issues for critics.

First, if literacy is introduced into the balance of world cultures, will this new technology alter the cultural identities of the nation-states affected? In our view, based on all of the prior studies examining the transition from orality to literacy, the answer is decidedly "yes." We are likewise convinced that literacy cannot function as a secondary option or technique in human communication. As we suggested earlier in this chapter, the extensive form of schooling involved in acquiring literacy encourages cognitive and sociological changes. A decision to encourage worldwide literacy may also involve a commitment to cultural transformation on a global scale.

Second, literacy frequently embodies an attitude of superiority that can demean oral culture members. Literacy involves the understanding and mastery of communication technologies that are external to the human being, and that require unique kinds of intelligences to use effectively (Gardner, 1983). Not everyone can master literacy equally. As is true of any symbolic system that distinguishes and classifies individuals, literacy differentials can be used to rank individuals and attribute greater intelligence, insight, and power to those who have more effectively mastered the literacy. If dominant and subordinant relationships are created among people based on their ability to deal with writing and print, literacy becomes a "political" issue. In this view, literary elites have maintained that literate cultures are "better" than oral cultures. The ability to recognize class differences does not, of course, provide a warrant for a qualitative rank ordering of these unrelated class items. Indeed, from a methodological perspective, unrelated class items cannot be qualitatively ranked. In this context, we believe it is appropriate to reconsider educational systems that assume that the mastery of writing and print are "more important" than the mastery of oral communication skills. Additionally, when we note that poor school districts are more likely not to have computers, we believe it is appropriate to reconsider what is meant by, and what the motives are for the introduction of, terminologies such as "information discrimination" as the foundation for national policies. In all, we believe it is appropriate to reconsider any symbolic formulation that posits that one set of communication technologies is "better" than another.

Finally, the basic units of literacy itself need to be examined anew. The totalistic impact and political elitism of literacy require that the literate reexamine the mode of communication they use and prescribe for others. The most exacting and thoughtful analysis of the basic units of literacy should be undertaken. We believe such an exploration has been provided by Weaver (1953), when he linked the grammatical features of literacy to questions of ethics. Accordingly, we find Weaver's essay to be a significant and powerful method of examining some of the essential issues related to literacy and its political and ethical status. We use the model outlined in Chapter 3 to organize and summarize Weaver's analysis. However, we should first note that at key junctures in his analysis we disagree with Weaver, and we hope you will consider Weaver's arguments with tremendous care.

In a chapter entitled "Some Rhetorical Aspects of Grammatical Categories," in his volume *The Ethics of Rhetoric* (1953, pp. 115–142), Weaver provided a critical analysis grounded in the essential nature of a literate technology. The essay highlights the importance of the grammatical rules that govern language use in a literate culture. By featuring the grammatical rules of literacy, Weaver has endeavored to show how they function rhetorically to shape appropriate or ethical social conduct. Weaver has argued that the formal patterns of grammar sustain the "public character" of language and meaning, "and when one passes the outer limits of the agreement, one abandons comprehensibility" (p. 115). In sum, Weaver's chapter suggests that a consistent interpretation of a text is possible through a clearer understanding of the influence of grammar. This type of consistency highlights the literate transmission of cultural understandings from one generation to the next.

Weaver's chapter also reveals the degree of influence that media impose on a culture. He has noted at the outset that "our concern is primarily with spoken rhetoric, which cannot be disengaged from certain patterns or regularities of language" (p. 115). However, the balance of the chapter considers conventions more consistent with literacy than with orality. As a result, Weaver's words are simultaneously prophetic and self-revealing because he has implicitly acknowledged the pervasive incursion of literate conventions into oral constructions. Accordingly, "Some Rhetorical Aspects of Grammatical Categories" provides an appropriate critical lens for viewing literacy as a technocultural drama.

Structural Analysis

Weaver's (1953) consideration of "Some Rhetorical Aspects of Grammatical Categories" focuses on production components consistent with literate communication systems. Although he has not explicitly addressed visual mechanisms per se, Weaver has explored a number of concerns that highlight literacy's visual nature. For example, he has steadfastly maintained that grammar constitutes "a system of forms," or "formalizations of usage" that constrain rhetorical selection and arrangement (p. 115). As a result, "style shows through an accumulation of small particulars" (p. 116). By concentrating on the rhetorical powers of grammatical forms in general, Weaver has placed stylized forms of presentation, a primary characteristic of visual mechanisms, at the heart of his essay.

Indeed, Weaver has contended that through repetitious use the simple sentence initiates a pattern of experience that establishes or constrains meanings through its formal unity (pp. 117–118). In addition, he has held that a sentence's unity "comes to have an existence all its own" (p. 117). Similarly, the complex sentence forces the reader "farthest into the reality existing outside self" (p. 124). As a result, grammar may be reasonably cast as a phenomenon exterior to direct experience, another essential quality of visual communication mechanisms.

Moreover, Weaver has cast language itself "as a standard objective reality" (p. 116). As such, language reflects the abstractness inherent in literate communication systems. From Weaver's perspective, framing a sentence requires "analysis and re-synthesis," intellectual procedures that characterize a structural feature of literacy (p. 117). Further, Weaver has held that the simple sentence emphasizes the "discreteness of phenomena within the structural unity" of the sentence (p. 119). That is, the "pattern of subject–verb–object . . . leaves our attention fixed on the classes involved" (p. 119). The segmenting of reality and subsequent discrimination between segments imposed by grammatical rules aptly identifies abstract thought and language as a structural feature of great importance in the critical analysis of literacy.

Linear and sequential processing also receive attention in Weaver's essay. He has contended, for example, that the "single subject–predicate frame" of the simple sentence lists or itemizes phenomena, "and the list becomes what the sentence is about semantically" (p. 119). This sequential construction of the world and events is also evident in Weaver's consideration of complex sentences. In fact, he has argued that understanding a complex sentence requires discriminating activity that distinguishes co-existing classes "according to rank or value, or places them in an order of cause and effect" (p. 121). In this way, information is ordered spatially or causally, consistent with the demands of linear and sequential processing inherent in literate communication systems.

Weaver has also acknowledged the importance of text as context as a structural feature in the critical analysis of literacy. He has contended, for instance, that "the rhetoric of any piece is dependent upon its total intention" (p. 117). From this view, meanings are to be negotiated through the text itself. In addition, Weaver's general focus on grammatical categories acknowledges that the text supplies an awareness of literate traditions, linguistic conventions, and the ideological assumptions of a culture. He has argued, for example, that because patterns are often recognized before meanings are interpreted from words, that "word classification and word position cause us to look for meaning along certain lines" (p. 118). In this instance, the grammatical categories embodied by a text cue the reader to recognize appropriate literate traditions and linguistic conventions. Likewise, Weaver's consideration of compound sentences firmly grounds the ideological assumptions of a culture within the text. In effect, the compound sentence presents ideas in balance, or as a "tension of stasis" (p. 124). Weaver has contended that the compound sentence was widely used during the 18th century where "counterpoise . . . was one of the powerful motives of their culture" (p. 124).

Cognitive Analysis

Weaver's (1953) analysis is also consistent with the manner in which a literate communication system can be said to shape thought processes, meanings, and understandings. In fact, he has proposed that sentence form reflects a necessary psychological and logical "operation of the mind" (p. 117). This operation of the mind engages in analysis of discrete classes of information and their resynthesis into an understandable linear pattern, or sentence. More importantly, the emerging intellectual pattern affected by sentence structure directs a writer or reader to an appropriate interpretation of a text's substance (see, e.g., p. 118).

Weaver's distinctions between sentence types provide a clear example of how form influences perception. Although the simple sentence "tends to emphasize the discreteness of phenomena within the structural unity" of a sentence and represents an "unclouded perspective" (pp. 119–120), the ranking or valuing of information in a complex sentence suggests a reflective mind capable of distinguishing differences beyond simple perception. Weaver has suggested that the mental refinements necessary to employ or understand complex sentences might reveal a causal principle at work and an understanding of dependence. Together, these processes might manifest themselves as a spatial, moral, or causal hierarchical ordering that reflects "the critical process of subordination" (p. 121). Weaver has further contended that complex sentences encourage thought characterized by involution, "or the emergence of one detail out of another" (p. 122).

In his discussion of the compound sentence, Weaver has continued to expand his notion that literate codes shape intellectual processes. He has contended, for example, that the balance of forces in a compound sentence requires a mental capacity for rationalism, subjectivism, and humanism. Compound sentences, then, reflect an artistic equivocation that provides more reasoned completeness to an expression. In effect, the compound sentence corresponds to a "philosophical interpretation rather than with the factual reality" (p. 126).

In essence, Weaver's essay is designed to articulate and maintain an intellectual hierarchy. He regularly identifies grammatical proficiency as a source of power to facilitate or hinder change (see, e.g., pp. 119, 142).

However, because he has frequently tied grammatical proficiency to an intellectual elitism (pp. 121, 123, 143, 142), facilitating change is properly understood as reinforcing the status quo. As a result, Weaver's efforts are in direct opposition to a postmodern analysis. Throughout the essay, Weaver has attempted to limit the possible interpretations for the influence of grammatical categories. Indeed, he has worked steadily toward a universal, immutable perspective.

Despite the lack of alternative readings offered by Weaver, this level of analysis is informative because it identifies the degree to which literacy and literate conventions can constrain human intercourse. From Weaver's perspective, the intellectual elite are best suited for citizenship, governance, and decision making. He has argued that language proficiency empowers citizenship and decision making. But he has also noted that "the work is best carried on . . . by those who are aware that language must have some connection with the intelligential world" (p. 142).

Sociological Analysis

Implicitly, Weaver (1953) has suggested that grammatical proficiency sustains social processes indicative of literate societies. By articulating distinct, regular rhetorical aspects of grammatical categories, Weaver has made it clear that writing and reading are solitary activities. Without standardized categories and codified usage, arbitrary and individualized interpretations are possible. In effect, a set of grammatical standards ensures and sustains literacy itself, maintains a sense of community, and preserves and transmits cultural understandings.

In addition, Weaver's tendency to promote intellectual elitism would consolidate literacy among limited social groupings. By suggesting restricted access to literate communication systems, Weaver's perspective encourages monolithic social institutions that bureaucratize and institutionalize cultural practices and traditions. Academia, for instance, would certainly be sustained by a public acknowledgment of the literate superiority of faculty. Essentially, Weaver's perspective encourages a renewed monasticism where colleges and universities are the controlling forces of cultural knowledge and appropriate social conduct. Similarly, contemporary governments would increasingly reflect the status quo. In effect, written constitutions become social covenants for the endurance of governmental forms. Even broad philosophical or intellectual movements might evolve into social institutions as universal interpretative frameworks. For Weaver, science has followed this process. In his view, the complex sentence played an instrumental role in the development of science. He has argued that patient, disciplined observation and the presentation of the objective world as a series of details models the scientific mentality (pp. 122–123).

Evaluative Analysis

Weaver's evaluation of the rhetorical aspects of grammatical categories suggests that grammatical proficiency is a requirement for citizenship. Without "some communal sense of the fitness of a word or a construction for what has communal importance," people are ill-prepared for governance and decision making (p. 142). For Weaver, grammar is a hortatory force in literate societies, directing attention to appropriate message construction, accurate interpretation, and appropriate ethical standards for communal involvement.

| CONCLUSION

This chapter has identified the structural features of human communication in a literate culture in which visual perception, linear and sequential processing, text as context, and abstract thought and language have been featured as essential features of literacy as a communication technology. In addition, the cognitive implications of this mode of communication have been suggested. This chapter has identified several of the social institutions that have been generated and sustained by writing and print. The chapter concluded with an illustration of a media critique indicative of technocultural dramas constituted by a literate communication system.

CHAPTER 6   The Electronic
Culture

Today, most of us are almost unconscious of how frequently and intensively we use electronic technologies. With hundreds of channels on some cable systems, people, especially men (Carter, 1991, p. D6), enjoy switching from channel to channel in a process informally identified as "zapping" or "channel surfing." With remote controls in 80 percent of American homes (Nielsen Report, 1991), zapping has been increasingly recognized as an option for dealing with commercials and boring programs, or for comparing programs. Indeed, some 17.9 percent of viewers are "heavy zappers," switching channels more than once every 2 minutes, while an additional 35.8 percent are "moderate zappers," switching channels one to three times every 6½ minutes. With time, zapping may become more selective (see, e.g., Eastman & Newton, 1995), but it continues to remain an important way of dealing with television programming.

Whenever zapping occurs, mass communication systems undergo a transformational change. Mass communication systems, as we have noted earlier, are traditionally conceived to be a one-way transmission from the producer (as *sender*) of the television program to the television viewer (as *receiver*). In this traditional formulation, the *sender* and *receiver* labels reflect the belief that the television producer or sender creates and determines the nature of the message conveyed to the television viewer or receiver.

But zapping changes these labels dramatically. While zapping, the receiver is creating a televised message. If allowed, and when in the mood, thanks to the marvels of the remote control, a viewer might even be able to enjoy watching four to five different television programs at one time. Admittedly, this kind of zapper misses the majority of any one of these programs being transmitted. However, from a creative perspective, the zapper is actually creating a new program (from the combination of other programs), in the sense that the zapper is watching a combination of electronic signals no one else is likely to be watching. In this way, the zapper changes the traditional role of the viewer from passive receiver to active programmer of the array of video signals transmitted by producers. In other words, the role of the passive receiver is becoming increasingly interactive. In addition, of course, interactive media (e.g., virtual reality and cyberspace systems) are an emerging class of media electronics.

Indeed, the electronic media are undergoing rapid and amazing transformations. In this sense, zapping is only one example of the new classes and new uses that can be made of electronic communication systems. In many respects, the traditional telecommunication systems (e.g., television and radio) are undergoing radical changes in which viewers become the central agent controlling and determining the message created by the electronic media (i.e., interactive media). In all, we anticipate that telecommunications and interactive electronic technological systems will continue to develop and to merge.

This chapter focuses on social systems that sustain cultural values and lifestyles from one generation to the next through electronic media. Specifically, this chapter explores the merging of electronic media, identifies the structural features of electronic media, examines the sociocultural influences of electronic media, and considers a critical example of interactive communication. In essence, this chapter investigates the role of telecommunication and interactive communication systems in constructing electronic technocultural dramas.

THE EMERGING AND MERGING ELECTRONIC MEDIA

In an electronic culture, telecommunication and interactive communication technologies constitute the web unifying the thoughts, speech, action, artifacts, and knowledge transmitted from one generation to the next. Specifically, any number of media have been formally classified as electronic, including the telegraph, telephone, film, television, radio, fax, tape recorder, and computer. Ultimately, a communication technology may be cast as an electronic medium when it transmits electrically created messages "almost instantaneously from a distance, eclectically combining oral and written forms of mediation with facsimiles of primary sensory encounters, also storable, and often readily malleable" (Gozzi & Haynes, 1992, p. 6).

Yet given the constant innovations in new electronic communication systems, the sheer number and variety of electronic media invite the creation of analytical categories and, at the same time, defy such categorization. However, Rogers (1986) has developed an appropriate analytical framework for understanding electronic media. He has cast electronic media as either one of two types: telecommunication or interactive communication. Proceeding chronologically, Rogers (1986) has isolated a telecommunication "era" that roughly spans the years 1844 to 1945. Based on the common functions of communication technologies emerging in this period, he has defined telecommunication as "communication at a distance . . . instead of moving people to ideas, telecommunication moves ideas to people. . . . Several of the most important telecommunication technologies (e.g., radio, film, and television) were primarily one-way, one-to-many mass media" (p. 28). He also includes the telephone and telegraph, primarily one-to-one and somewhat interactive technologies, in his category of telecommunication technologies.

Rogers's era of interactive communication begins with the 1946 development of ENIAC, the first mainframe computer, and includes currently emerging two-way or interactive communication technologies. Interactive communication behaviors, Rogers (1986, pp. 30–33) has reported, "require a high degree of individual involvement" that can be provided by any number of two-way media, including microcomputers, teleconferencing, teletext, videotext, interactive cable television, and communication satellites.

In all, telecommunications and interactive technologies characterize the two dominant types of communication technologies creating the electronic culture. We suspect that these two forms are undergoing transformations in which they increasingly affect one another.

This transformation is understandable and predictable in two important ways. First, the communication technologies of each are essentially electric ones. Although a case could be made for treating telecommunications and interactive technologies in separate chapters, both focus on how individuals experience understanding and communication in an electronic culture. Chesebro and Bonsall (1989) for instance, have argued that "virtually all of the new information technologies introduced within the last 50 years have relied upon electricity as a central component in their design" (p. 15). Because electricity acts as a central component in the design of new technologies, telecommunications and interactive technologies literally share an "epistemological zone" (Gozzi & Haynes, 1992, p. 2) where cultural knowledge, values, and identities are shaped and preserved.

Second, telecommunications and interactive technologies are merged in this chapter because the specific technologies that distinguish the two are now merging. Rogers (1986) has noted, "The new electronic technologies are causing an integration of media that we have conventionally considered to be completely separate" (p. 31), a view shared by a host of other scholars (e.g., Carey, 1989; Chesebro & Bonsall, 1989; Meyrowitz, 1985; Rheingold, 1991; Williams, 1991). Indeed, Williams (1991) has particularly maintained that "as telecommunications and computing progress and coalesce, they are blending into a single system which many are now calling the 'intelligent network' " (p. 8).

Five examples of current electronic technologies appropriately highlight and illustrate the merger between telecommunication and interactive communication systems.

First, as we indicated at the outset of this chapter, the remote control can blur the distinction between the two systems. Many homes are equipped with several remote controls, for use with the television, VCR, compact disk system, stereo system, and/or satellite system. Users of the television remote control often engage in zapping, frequently switching channels to edit their television viewing. Zapping is an important cultural practice because it underscores how people can use one minor technology to create their own programs. Viewers change channels so frequently that they begin watching a "program" they have made up through their zapping. The use of the television remote control ultimately allows the viewer to merge advanced computer technology with the television, thereby creating a semi-interactive communication system.

Second, electronic technologies are becoming more specialized. Cable and satellite reception systems have increasingly shifted electronic technologies from a mass- to individual-oriented medium. Technology has become so sophisticated that it increasingly responds to individual rather than mass, cultural, or even group preferences. Indeed, as Rogers (1986) has so aptly maintained, the new media have become "*de-massified*, to the degree that a special message can be exchanged with each individual in a large audience" (p. 5). For example, the increase in channels, from a mere handful in commercial television to the hundreds available through cable and satellite systems, allows television viewers to select programming from a seemingly infinitely diverse menu. This expanded television menu provides audiences with the opportunity to make unique viewing choices according to their own particular interests. In this way, the expanded television menu serves an interactive function.

The use of videotapes on television systems provides a third example of the current merger of telecommunication and interactive communication systems. Videotapes allow viewers to make, literally, their own programs, and they can also watch them whenever they want. The zeal with which American audiences have incorporated videotape and videodisc systems into their lives suggests that the degree of interactivity provided through these systems is compelling. It is not surprising then, that Dizard (1989) has reported that "over 50 million United States homes were equipped with video recorders by 1988" and that "Americans bought tens of millions of blank tapes annually" (p. 122). By 1990, the number of VCRs in the United States had grown to sixty million, with ten million new units being sold each year, and with more than 40 percent of American homes enrolling the VCR in their entertainment system (Corrigan, 1991).

Fourth, advances in computer technology continue to generate opportunities for users to construct communication environments. A video game called *Punch Out* provides an early example of the blurring of source and receiver in interactional systems. Users select an opponent with specific racial or ethnic traits from a list of possible opponents. Then the user tries to electronically batter the opponent into submission. Certainly, the ideological implications of this particular video game should not be overlooked. Nevertheless, the interactive potential of the game is clear. Compact disc read only memory (CD ROM) programs offer a similar example. Databases such as *Collier's Interactive Encyclopedia* can be entered at different points. As a result, users are not confined by the strict linearity that literate systems might impose. No two constructions based on CD ROM data are likely to be the same. In this environment, new ideas, concepts, and programs are possible.

Fifth, and finally, telecommunication and interactive systems have increasingly merged because audiences are increasingly becoming active agents when they deal with media systems. Investigations of the relationship between audience and mass media suggest that the content of television programming is essentially polysemic, open to a variety of individual interpretations (see, e.g., Fiske, 1986). For some, the existence of polysemy highlights the idiosyncratic, which makes it unlikely that a culture can be created or sustained. However, the polysemic nature of television texts is grounded in a multicultural social structure. Liebes (1988) has demonstrated that interpretive communities exist wherein audiences actively negotiate their readings of television programs based on the values and narrative forms of their community. Similarly, Condit (1989) has argued that audiences actively shape their readings of television texts within contextual constraints. In fact, the notion of active audiences constructing their own interpretations of texts within the constraints of a cultural system has become a commonly accepted viewpoint among communication researchers (see, e.g., Biltereyst, 1991; Cohen, 1991; Grodin, 1991; Hacker & Coste, 1992; McDonald & Schechter, 1988; Press, 1991b; Sholle, 1991; Shotter & Gergen, 1994).

The blurring of distinctions between telecommunications and interactive technologies marks an increasing shift from information acquisition to knowledge construction. In electronic cultures, receivers are increasingly called on to create messages from the world around them. More and more, the job of message construction falls to the receiver (see, e.g., Rogers, 1986). As new communication technologies continue to be developed, an individual technology's power will rest in its ability to consolidate existing technologies to facilitate message construction (see, e.g., Williams, 1991).

THE NATURE OF INFORMATION IN AN ELECTRONIC CULTURE

Electronic communication systems incorporate a variety of structural features. Watching television, for instance, requires an ability to use and interpret audio, visual, and print systems all at once. For a viewer to appropriately interpret an electronic message and construct a corresponding reading of that message, some knowledge of sound, nonverbal behavior, imagery, and linear progression are often necessary. Watching an early newsreel, for example, required an audience skilled in constructing a message from a series of newspaper headlines interspersed with brief film sequences and supplemented by dialogue and background music. In similar fashion, using a personal computer to write a book requires an awareness of literate conventions as well as some understanding of the technologies and software required to access different information and formatting systems. Obviously, an appreciation for some oral and/or literate structural features is often a prerequisite for comprehending electronic messages. For example, Eckhardt, Wood, and Jacobvitz (1991), have maintained that "verbal ability and prior knowledge play important roles in the memory for and comprehension of complex televised narratives" (p. 645).

The structure of a specific telecommunication system or interactive communication system would include some combination of audio, visual, print, and information system technologies. However, electricity constitutes the unifying structural feature of all electronic technologies. A number of scientific developments contribute to advances in electronic technologies, but chief among them would be the discovery of electricity and its application to human communication.

Electricity

Conversion of natural resources to electrical energy combines generating, storing, and transmitting systems. These systems were developed through the cumulative and collective efforts of thousands of philosophers, scientists, inventors, and technicians over hundreds of years. The history and study of electricity can be traced at least to the Greek philosopher Thales in 600 B.C. In fact, the word "electricity" derives from the Greek word for "amber," which when rubbed, Thales discovered, can attract small, light objects (Asimov, 1963, p. 114). Asimov has reported that 18th-century scientists provided a significant breakthrough when they developed the Leyden jar that concentrated electricity and could generate a spark. Popularized by Benjamin Franklin in 1752, electricity was introduced to the study of anatomy by Luigi Giovani in the 1780s. While investigating Giovani's discoveries, Volta constructed the first electric battery in 1800 (Asimov, 1963, p. 115).

The study of electricity has been closely tied to the study of the human body. Ancient and medieval physicians believed that a fluid flowed through the nerves in the human body, but no observational basis existed to characterize or study the effects of this fluid. The discovery of electricity, however, allowed scientists to approximate mind–brain activity. Moreover, the 20th-century interest in the atom and its structure provided a language and model for understanding the incredibly complex activity of the human brain (Asimov, 1963, pp. 115–116).

In this context, it is therefore not surprising that human communication has been cast as a mind–brain activity. Indeed, Gregg (1984) has maintained that "all mind–brain activity is symbolic. Put another way, all human experiencing is the result of brain processing which creates the structures we call 'meaning' " (p. 17). In his view, the brain processing occurs electronically, complete with neurons, neuron firings, and the capacity of the human brain to "rewire portions of itself when conditions seem to warrant" (p. 18). Electricity, then, is not simply a structural feature of telecommunication systems and interactive communication systems, but it can be understood as pervading cultural conceptions about mind–brain activity, cognitive processing, and the symbolic structuring of reality.

In all, the discovery of electricity and its subsequent social applications transformed human conceptions of the mind as well as the ways in which humans communicated with one another. Put simply, electricity reconfigured the way humans understand and talk about their world.

As the basic structural feature of electronic cultures, electricity makes it possible to combine a variety of other structural features. This combination of features defines the limits of the speed of telecommunication and interactive communication systems (see, e.g., Schlater, 1970). Indeed, telecommunication and interactive communication systems are increasing and will continue to increase information processing speeds (see, e.g., Rheingold, 1991; Roszak, 1986; Williams, 1991; Zuboff, 1988). To facilitate high-speed message construction, electronic communication technologies employ a combination of structural features from among audio, visual, and print.

Audio

In one of the most classical conceptions of media, McLuhan (1962, 1964) has held that the media are properly perceived as extensions of human organs and abilities. From this perspective, media augment human reach by amplifying one or more sensory modes. In this view, sound is electrically replicated through an audio system. Telecommunication and interactive communication systems often employ an audio system in combination with other structural features. These auditory systems are constituted by basic auditory units, storage and transmission technologies, and auditory forms. A full discussion of auditory systems might appropriately include a detailed section about the broad technical components of sound production systems. However, the present project forgoes a technical discussion because it is primarily concerned with isolating the general production features of auditory systems.

The basic auditory units of an electronic audio system include amplitude (loudness), frequency (pitch), attenuation (duration), quality, harmonics (overtones), and interference. Each of these units are produced or occur during human speech (see, e.g., Eisenson, 1974; Lessac, 1967; Zemlin, 1981). Early speech and hearing scientists from the 13th through the 19th centuries experimented with the development of human sounds, often creating elaborate machines that attempted to replicate human sound (Zemlin, 1981). In many ways, the current technologies of audio systems are the electrical offspring of these early sound machines. However, when produced electrically these basic auditory units may be manipulated and recombined into unique forms that often exceed the capabilities of the human voice.

The control and manipulation of basic auditory units into audio messages require the use of storage and transmission technologies. Audio systems commonly employ recording technologies such as discs, tapes, and channels to store audio information. Once recorded, audio messages can be accessed through any number of receiving technologies: radio, stereo, television, telephone, and computer (see, e.g., Sterling & Kittross, 1978). Audio systems may exist as autonomous media like radio or stereo, or as a component of other media, such as television, film, or computer. Audio recording and receiving technologies make it possible to access, manipulate, and recombine audio information in unique and unexpected ways.

One of the most frequent applications of audio systems occurs when sound is added to visual images. Ever since sound was introduced to Hollywood movies in the early 1920s, television and film have relied heavily on audio systems to enhance storytelling (see, e.g., Bordwell, Staiger, & Thompson, 1985; Zettle, 1973). Audio systems employed in television and film generate a variety of auditory forms that are recorded and stored as separate dialogue, music, and effects soundtracks and mixed to produce the audio portions of television and film. Although emerging interactive technologies often employ audio systems in new and unique ways, their audio systems rely on the same functional tendencies as those utilized in television and film. Accordingly, several audio forms are appropriately considered here.

Dialogue, or characters saying words to other characters, is a common feature of television and film. Affron (1982) has suggested that dialogue encourages viewers to construct an affective relationship with the narrative events of a film. As a result, dialogue promotes narrative continuity that helps viewers follow a film's story. Similarly, voice-over narration, spoken language directed to the audience, serves the narrative activity of the film by establishing perspective, or point of view, and also by identifying the narrator as an imaginary witness to story events (see, e.g., Bordwell, 1985). Spoken language in film and television is stored as a dialogue soundtrack.

Music soundtracks may include local and background music. Local music is music that originates in the scene or shot, as for example, when an actor in a musical breaks into song, or when a stereo or band provides dance music at a party. Background music on the other hand, is music without any identifiable source in the scene. Hollywood films and television programs rely heavily on background music to maintain narrative continuity (Bordwell, 1985; Bordwell et al., 1985). The entry of the evil villain in a film is often preceded by a sinister or frightening instrumental piece. Likewise, a romantic interlude can be introduced with a love song. In both cases, music engages the audience at an affective level and invites viewers to align themselves with a story's events (Affron, 1982).

Sound effects, the noises that naturally accompany a scene, round out the soundtrack in film and television. Early sound effects can be traced to the emergence of radio drama. The ambient sounds, or naturally occurring sounds of a scene, were often produced in the radio studio using artificial devices. Later, these sounds were recorded, catalogued, and played whenever the drama required them (see, e.g., Sterling & Kittross, 1978). The same practice is followed in television and film. Like dialogue and music soundtracks, the effects soundtrack helps the viewer or listener sustain narrative continuity.

Interestingly, the absence of sound serves a similar function. Silence or the disruption of ambient sound, in a film or television program often directs the viewers' attention to a visual cue, or affords the viewer an opportunity to more closely study the visual presentation. In this way, silence contributes narrative continuity to visual discontinuity (see, e.g, Bordwell et al., 1985; Zettle, 1984).

Basic auditory units, storage and transmission technologies, and audio forms constitute the essential features of auditory systems. The primary function of an auditory system is to replicate, as closely as possible, the real time and space of human interaction (see, e.g., Affron, 1982; Bordwell, 1985; Doane, 1985). As an autonomous medium or as a component of a more integrative medium, audio systems encourage a fiction of human presence (see, e.g., Affron, 1982). Audio systems also allow viewers to follow and coordinate visual constructions.

Visual

Telecommunication and interactive communication systems increasingly include a visual component. The visual component of any electronic medium is comprised of basic visual units, the composition of the visual unit, image size, perspective, camera speed, lighting, continuity, laboratory devices, animation, graphics, color, and recording and receiving technologies.

Like audio, visual features seem designed to replicate some parts of human interaction or to engender a sense of human activity and presence in electrically mediated messages. Both audio and visual components are regularly combined in some telecommunication systems and interactive communication systems to produce coherence and continuity of action from discrete and discontinuous images and sounds. Visual images contribute to the fiction of human action by generating cinematic depth, activity, and space (Affron, 1982, p. 139; Arnheim, 1974; Zettle, 1973). In other words, visual images redefine "the junctures and the distinctions between fictivity and verisimilitude," or blur the imminent margin between fiction and reality, thus providing a locus for enactment (Affron, 1982, pp. 133, 142).

The basic units of the visual component that begin to articulate cinematic or imagistic activity and presence include the frame, the shot, the scene, and the sequence. A frame is a single photographic unit (see, e.g., Giannetti, 1976; Harrington, 1973; Monaco, 1981). When several frames are exposed continuously they constitute a shot, the single uninterrupted action of a camera (Salt, 1985; Sandro, 1985; Vaughn, 1985). Shots serve a variety of functions. For instance, shots may be employed to identify the location of a scene (extreme long shots or establishing shots), or to focus attention on a specific visual subject (close-up) (see, e.g., Balazs, 1970; Williams, 1965). A series of shots in one location or in the same apparent time period comprise a scene. Similarly, a series of interrelated scenes or shots unified by one location, time period, generating action, point of view, or character interactions constitutes a sequence (see, e.g., Kawin, 1987). Each of these basic visual units may be employed separately or in combination in telecommunication and interactive communication systems.

The basic units carry information by virtue of the composition of the frame and the movement within the frame. The composition of a frame commonly refers to the physical relationships between elements fixed within the image (Herbener, Tubergen, & Whitlow, 1979). The location of objects and people may be balanced, with proportionate distribution of elements and space, or frame composition may be geometrical, with elements organized in triangular, circular, or rectangular form (Kawin, 1987). The organization of the physical elements of a frame provides a superficial space, a hint of human presence.

To encourage a greater sense of activity and presence, visual units also include movement within the frame. Visual movement can be indicated through a variety of camera movements or shots (Bordwell et al., 1985; Giannetti, 1976; Kawin, 1987; Monaco, 1981; Wurtzel & Dominick, 1971). Movement across a scene, for example, can be suggested by pivoting the camera horizontally, or "panning" the scene. In contrast, placing the camera on a crane or boom can be used to promote vertical movement. A sense of visual movement and depth can also be generated through tracking shots, where a mobile camera follows the subject, or through dolly shots, where the camera is mounted on a platform with wheels. Camera angles and lenses may provide additional indications of movement within a frame (see, e.g., Harrington, 1973; Tiemens, 1970). Tilt shots, for instance, create added depth and movement by rotating a stationary camera at an angle, either up or down. Zoom lenses serve a similar function by moving the viewer closer to or further from the frame. Finally, the presence of movement in the composition of the frame can be intimated through the use of a library of stock shots, borrowing shots taken from another show, program, or film.

Image size can also be manipulated to influence perceptions of presence, depth, and activity. Image size is often determined by the distance between the camera and the object of the shot, or by the focal length or amount of area a lens can photograph from a given distance. In both instances, image size helps construct a perspective for viewing visual imagery (Bordwell, 1985; Zettle, 1973). Perspective, the relative positions and relationships of objects to the viewing eye, can be enhanced and adjusted through the use of different camera lenses (e.g., normal, wide-angle, telephoto, long, and fish-eye), camera filters, and camera placement (low- and high-angle shots) (see, e.g., Giannetti, 1976; Kawin, 1987; Monaco, 1981; Wurtzel & Dominick, 1971).

Camera speed and lighting also constitute important elements of a visual system. Slow and fast motion, for instance, can produce unanticipated and unfamiliar movement in a film. Slowing subjects' actions below normal rates of human activity or accelerating them beyond normal rates are regularly used to depict dreamlike states or to promote comedy (see, e.g., Giannetti, 1976; Schlater, 1970). Because camera speed essentially regulates visual action, viewers are invited to focus their attention on the grace of human action, or its folly.

Lighting, both natural and artificial, contributes to viewer perspective. Key lighting, which highlights a specific subject, and fill lighting, which complements key lighting by illuminating adjacent subjects from the front, side, back, top, or bottom, play an instrumental role in directing attention to or away from a specific subject (see, e.g., Harrington, 1973; Millerson, 1982; Monaco, 1981).

Other visual production elements sustain the continuity of the film. The visual constructions of telecommunication and interactive communication systems are often nonlinear. The shots included in a sequence may be made at different times and at different locations. Therefore, suturing discrete visual images and sounds into a narrative whole demands a technological means of disguising the discontinuous aspects of film (see, e.g., Bordwell, 1985; Wurtzel, 1983). Cinematic continuity can be provided through dissolves, wipes, flashes, flips, split-screens, freeze-frames, fades, montage sequences, and jump shots. Laboratory devices like double exposure and negative images can also be added to enhance visual imagery. In addition, color, animation, and graphics can be employed to transform the visual image (Bordwell et al., 1985).

Like audio systems, visual systems may be accessed through a variety of storage and transmission technologies. Visual images may be recorded and transmitted live, or they may be recorded and stored for later showing on film, videotape, and disk. There are also a variety of technologies employed to receive visual images, including televisions, movie screens, and computer monitors. However, the reception of visual imagery is shaped as much by the screen size and the social setting as it is by the specific receiving technology. Viewing a film on a large screen in a theater is a far more public experience then viewing a television program in the bedroom. The intimacy of the small screen and private setting constitute a viewing environment very different from that of watching film at a theater. In this way, the viewing environment itself can be considered part of the visual apparatus of telecommunication and interactive communication systems.

Together, the production components of visual images comprise a significant structural feature of telecommunication and interactive communication systems. The visual component, often working in harmony with an audio component, contributes to the fiction of human action and presence. However, these systems may often be augmented by yet another structural feature, print.

Print

Audio and visual components seem to dominate telecommunication and interactive communication systems. However, print technologies continue to play an important role in electronically constructed messages. Print may emerge as the text of a telegram, as a newspaper headline in a film, as part of a television advertisement, as an E-mail message, or as videotext information. In many ways, the print mode used in telecommunication and interactive communication systems shares the structural features of print and literacy in general. It standardizes the text, fixes the text and its meanings, isolates the writer and reader, promotes consistency in interpretations, makes messages more accessible, reinforces the visual mode, and promotes cross-referencing (i.e., hypertext programs; see Bolter, 1991; "English Assignments," 1989; Ferrell, 1988; Gelernter, 1994a, 1994b; Grafton & Permaloff, 1991; Lanham, 1993; Muth, 1991; Perez, 1992).

| KNOWING IN
| AN ELECTRONIC CULTURE

Although audio, visual, and print—as information-accessing technologies—constitute the basic structural features of electronic communication systems, the individual information user or consumer functions as the creative core for integrating these modes of information into a coherent knowledge system with corresponding ways of perceiving and understanding human existence. With increasing regularity, researchers are casting human agents as a central structural feature of electronic communication systems.

Early mass communication theories hypothesized a passive audience that could be manipulated and persuaded by the immense power of the media. Although theories of this sort are known by a variety of names, such as hypodermic needle or transmission belt, they are commonly recognized as a variant of the magic bullet theory (see, e.g., DeFleur & Ball-Rokeach, 1982). A common assumption of the magic bullet theory is that media messages are uniformly received by all audience members and instigate immediate and direct responses.

However, as media research developed in the post-World War II years, empirical findings began to contradict the assumption of a passive audience. Increasingly, researchers began to recognize the active nature of audiences. Olson and Bruner (1974), for instance, have proposed that humans actively construct their realities and their representations of that reality. They have held that "living systems have an integrity of their own;

they have commerce with the environment on their own terms, selecting from the environment and building representations of this environment as required for the survival and fulfillment of the individual and the species. It follows that our conception of physical reality is itself achieved by selective mediation" (p. 128). They have further contended that "objects and events are not passively recorded or copied, but, rather, acted upon and perceived in terms of action performed" (p. 128). Similarly, Gross (1974) has argued that human competencies with social and intellectual skills are tied to mastery of those skills in the context of modes of symbolic processing (lexical, social–gestural, iconic, logical–mathematical, and musical). That is, he has suggested that developing the social and intellectual skills essential to exist in a specific culture requires active performance (or attempts at performing) within those symbolic processing modes.

More recent conceptions overtly recognize the active role of individuals in constructing messages and, concurrently, knowledge from their mediated environments. Potter's (1988) survey of media effects research, for example, challenges the assumption of passive audiences and identifies the importance of active individual interpretation in constructing reality from televised messages (see also Fiske, 1986). Likewise, Levy (1985) has explored the active audience concept in relation to VCR use. He has demonstrated that "VCR owners are essentially an active audience, that in general their orientations to VCR use are selective, somewhat involving, and often useful" (p. 273). Similarly, Turkle (1984) has emphasized that the computer invites users to actively construct, manipulate, and explore individual conceptualizations of reality. In her words, the computer is "an ideal medium for the construction of a wide variety of private worlds and, through them, for self-exploration" (p. 15).

Clearly, researchers who investigate electronic media have increasingly recognized the role of the individual in integrating information systems and actively constructing meanings. Yet Turkle (1984) has warned that computers can "trap people into an infatuation with control, with building one's own private world" (p. 19). From this psychological perspective, individual constructions of messages and meanings tend to be idiosyncratic. More importantly, as we noted earlier in this chapter, such idiosyncratic interpretations of the world and its events are unlikely to sustain a culture's values and lifestyles from one generation to the next.

However, as we have suggested earlier, when individuals provide interpretations of events, they are culture-bound (see, e.g., Biltereyst, 1991; Cohen, 1991; Condit, 1989; Grodin, 1991; Hacker & Coste, 1992; Liebes, 1988; McDonald & Schechter, 1988; Press, 1991b; Sholle, 1991; Shotter & Gergen, 1994). In this view, individual interpretations of mediated information or constructions of media messages are grounded in interpretive communities (Liebes, 1988), such as context or culture (Biltereyst, 1991; Condit, 1989), economic classes (Press, 1991b), or, more broadly, in a relevant synthesis of sociopsychological, textual, and conversational resources (Cohen, 1991; Shotter & Gergen, 1994). In sum, information users are active, and their interpretations of texts or constructions of messages are constrained by their cultural system. In other words, the individual becomes increasingly responsible for transforming random information into what that person's culture would recognize as knowledge. To this we would add that the particular medium or media employed to integrate information systems into a coherent message or knowledge base also constrains the available choices.

Several studies indicate that the notion of active individuals, actively constructing meanings from the events in the world around them and constrained by their cultural context, differs from media system to media system (see, e.g., Chesebro, 1989; Gozzi & Haynes, 1992; Havelock, 1986; McLuhan & Powers, 1989; Meyrowitz, 1985; Olson & Torrance, 1991; Ong, 1982; Pfau, 1990; Postman, 1985). In oral cultures, social constructions of reality are ultimately based on direct experience in the society's or group's oral traditions. As a result, oral cultures develop a community reality, shaped, shared, and sustained by the ongoing interactions of its members. In contrast, members of literate cultures construct a social reality through the abstract manipulation of information contained in symbolically coded texts. Literate cultures thus encourage an individual reality, where individual interpretation of written or printed information is tempered by the linguistic conventions and ideological assumptions inherent in the text.

These five structural variables of electronic cultures—electricity, audio, visual, print, and the individual—come together in yet another social construction of reality. In an electronic culture, the sociocultural influences of electronic media attest to the construction of a *virtual reality*. Accordingly, we turn our attention to virtual reality and its implications for life in an electronic age.

THE SOCIOLOGY OF THE ELECTRONIC CULTURE: SOCIAL INSTITUTIONS AND INDIVIDUAL EXPERIENCES

The concept of *virtual reality* or *virtual environments* has steadily emerged through the integration and sophistication of the electronic culture's structural variables. Because these variables may be combined in a variety of ways that are often unpredictable, researchers have been hard pressed to formulate a specific, common definition of the concept (see, e.g., Biocca, 1992a; Rheingold, 1991; Steuer, 1992). Steuer (1992) has reported that many of the popular definitions for virtual reality cast it as "a collection of technological hardware, including computers, head-mounted displays, headphones, and motion-sensing gloves" (p. 73). As a result, virtual reality can lack theoretical utility. Accordingly, Steuer (1992) has argued that "hardware" definitions encourage arbitrary classification of technological systems based on the presence or absence of certain technological components. Similarly, Biocca (1992a) has argued that when defined simply as "a class of computer-controlled, multi-sensory communication technologies," the concept of virtual reality encompasses a theoretical terrain as diverse as the possible technological combinations (p. 5).

Ultimately, there is no easy way to explicate the nature of a virtual reality. The issues involved in this process are richly illustrated in the analysis provided by Steuer, and we can gain an appreciation for these issues by briefly summarizing Steuer's analysis.

In his scheme, Steuer (1992) has proposed an experiential or psychological definition that casts the individual as the primary unit of analysis of virtual reality. He has specifically held that virtual reality can be appropriately defined by the concept of *presence*, the perception of the physical environment or "surroundings as mediated by both automatic and controlled mental processes" (p. 75). In his view, human perception of the physical environment is a direct experience, occurs naturally, and often goes unnoticed. However, he has argued that electronic technologies create a second, mediated environment. When the feeling of presence in the mediated environment exceeds the feeling of presence in the natural environment, Steuer has suggested that human presence may be cast as *telepresence*. Steuer argued the case by first considering what is meant by "presence." In his view, "*presence* refers to the *natural* perception of an environment, and *telepresence* refers to the *mediated* perception of an environment" (p. 76). In Steuer's view, "This environment can be either a temporally or spatially distant *real* environment (for instance, a distant space viewed through a video camera), or an animated but nonexistent *virtual world* synthesized by a computer (for instance, the animated world created in a video game)" (p. 76). Based on a broad application of the concept of telepresence, Steuer has defined virtual reality as "*a real or simulated environment in which a perceiver experiences telepresence*" (pp. 76–77).

Furthermore, Steuer has argued that virtual reality may be explored across two dimensions that determine telepresence in communication media: vividness and interactivity (see especially pp. 79–89). Vividness refers to the representational quality or sensory appeal of a mediated environment and is constituted by the breadth or number of simultaneous sensory dimensions present in a mediated environment, combined with the depth or quality of sensory information available in a mediated environment. Interactivity refers to the degree to which users may transform or modify the content and form of a mediated environment. Steuer has proposed that the interactivity of a given communication medium depends in large part on the speed at which the mediated environment can assimilate information, the number or range of possible alternatives for action within the mediated environment, and the degree to which a mediated environment maps human actions or corresponds to human actions in real time. Indeed, Steuer has speculated that "the more vivid and the more interactive a particular environment, the greater the sense of presence evoked by that environment" (p. 89).

In this regard, Biocca (1992a, 1992b) has implicitly held that interactivity and vividness contribute to a virtual reality consistent with the degree to which the components of a medium "match the requirements of our perceptual system" (1992a, p. 13). More specifically, Jones (1993) has argued that "the goal of virtual reality technology is creation of space, or, perhaps more pointedly, the control of the perception of space" (p. 246). Furthermore, he has argued for more emphasis on the aural dimensions of virtual reality. From his perspective, most conceptualizations of virtual reality feature the visual senses and overlook the significance of the aural dimension. In this way, he shares Steuer's and Biocca's conception of virtual reality.

From their viewpoint and, we might add, from our viewpoint, virtual reality initially should be viewed as a replication of real time and space by closely simulating human perceptual modes. Yet the concepts of the *replication* and *simulation* of reality and human perception remain problematic. Replications and simulations are generally viewed as polar opposites of what reality is and what the human perception process involves. However, the notion of a *virtual reality* constructs these paradoxical concepts as seamless.

The nature of a virtual reality more clearly emerges when it is compared to the kind of realities generated by other communication technologies. Four differences are particularly noteworthy.

First, the virtual reality created by electronic communication systems is intertextual. Drawing on the works of Kristeva, Hawkes (1977, p. 144) has defined intertextuality as the dependence of every text on other texts. Where literate cultures envision texts as autonomous and as the context for interpretation, electronic cultures embrace texts as social constructions developed from and constrained by prior cultural knowledge and experience, and constituted by messages from a combination of media systems (see also Bloome & Egan-Robinson, 1993; Chesebro, 1989; Shirane, 1990).

In electronic cultures, the meaning of a context shifts from the text as context to the medium itself as context. As a result, the intertextual nature of electronic communication systems encourages what Chesebro (1989) has called a *media context*:

> This media context might be defined, for example, by the barrage of 1,600 electronic advertisements on radios and televisions which invade, and often define a critical dimension of, an individual's environment. This media context presumes that an ongoing stream or set of messages are constantly at least part of the background, if not a primary set of environmental cues for understanding. (p. 9)

Intertextuality is a popular concept in electronic cultures. Much of contemporary literary critical theory, for example, is informed by notions of intertextuality (see, e.g., Durey, 1991; Toyama, 1990). Similarly, intertextuality has gained increasing influence in educational psychology. Intertextuality has been tied to knowledge assembly (Hartman, 1991), critical thinking (Short, 1986), and to defining and sustaining cultural values, lifestyles, and knowledge (Bloome & Egan-Robinson, 1993). Moreover, intertextuality regularly manifests itself in everyday environments. Chesebro (1989, pp. 7–8) has identified the intertextual nature of electronic communication systems in the language used to describe media interconnections, in the content of new technologies, and in emerging communicative forms. He has suggested that the episodes of a television show, film sequels, and the use of a popular song as the background for a film or television show all exemplify the intertextual nature of electronic communication systems (see also Olson, 1987).

By playing different texts off one another, intertextuality challenges prior explanatory narratives and generates new textual readings that lead to new knowledge and understandings. Ultimately, intertextuality helps account for the individual actively constructing "a discrete and independent social reality" (Chesebro, 1989, p. 9). Insofar as these individual realities lack the direct experiential base of orality, they may be properly conceived of as virtual realities. Ultimately, however, virtual realities must be understood as a fiction of phenomenal reality.

Second, because virtual realities are often discrete and independent, electronic cultures encourage a distinctive form of social organization. In electronic cultures, discrete and independent reality constructions invite an empathetic social organization instead of an experiential one. Gergen (1991), for example, has argued that contemporary social movements emphasize relational realities and juxtapose sympathetic alignments for direct social and political action. A similar notion can be discovered in Grossberg's (1986, 1988) concept of affective alliances. Gozzi and

Haynes's (1992) "empathy at a distance" and Bennett's (1992) "communities of the mind" also reflect the emergence of an electronic organization of society and culture. Not surprisingly, postmodern philosophy, which directly corresponds to the electronic age, specifically argues for a system of bloc politics that replaces broad political consensus with temporary covenants (see, e.g., Aronowitz, 1988; Bove, 1986; Lyotard, 1979).

In this view, social groupings, indeed, entire cultures may be formed on the basis of a shared construction of reality or a shared emotional commitment to a particular issue or interest. For example, the environmental movement encompasses a wide variety of specific interests. Environmentalists may be chiefly concerned with rain forest depletion, air and water pollution, or the increasing human threat to other life-forms. The specific interests of each group could be united in common cause in an electronic culture when faced with a natural disaster like a volcanic eruption, or when confronted with humans destroying pristine environments to search for gold. Networking by fax or by some combination of computer, telephone, and modem, environmentalists may share information, organize disaster response teams, and plan political strategies. The shared sympathy for the environment in general would overcome the independent and discrete constructions of reality of the groups and individuals involved. Moreover, electronic media make it possible for many to be directly involved without actually having any direct contact with the exigent event or the implementation of corresponding corrective actions. Virtual realities can then, in some ways, organize a culture's social systems. However, social and political organization in electronic cultures constitute what Edelman (1988) has called a "spectacle," where activism is supplanted by a socially constructed reality without an experiential context, where individual involvement can be restricted to selective levels of engagement.

Third, virtual realities can define and sustain cultural values, lifestyles, and beliefs. The socially constructed realities that develop from the transactions between film or television content and an active, interpretive audience, for example, are commonly understood to shape attitudes, behaviors, and ideologies (see, e.g., Chesebro, 1979, 1982, 1986a, 1991). Marc (1984) has argued that the combined content of American television programming constitutes a collective cultural consciousness. From this common repository of information, audiences extract frames of references for deciding how to act in different social situations (Biskind, 1983); determining what to think about people from different cultural, racial, and ethnic backgrounds (see, e.g., Bogle, 1989); understanding how ideologies emerge and infiltrate a culture (Nichols, 1981); recognizing a social value system (Quart & Auster, 1984); and appreciating the influence of film and television on learning and child development (see, e.g., Liebert, Sprafkin, & Davidson, 1982).

The more interactive communication systems also sustain a view of appropriate social conduct and behavior that is directly related to the knowledge and skills necessary to be a proficient user of the medium. A special issue of *Communication Education* (Vol. 43, April 1994), for example, is devoted to extolling the potential of the National Information Infrastructure (NII) and invites readers to learn to use the NII or be left behind in the search for information in the future (see Phillips, 1994). Contributing authors provide terminology (Santoro, 1994), instructions (Berge, 1994; Collins, 1994; Rowland, 1994; Ryan, 1994), educational applications (Bailey & Cotlar, 1994; McComb, 1994), business and industrial applications (Barnes & Greller, 1994), network resources (Benson, 1994), and a research agenda (Kuehn, 1994). In combination, these essays trace the cultural landscape to be found in the virtual reality of the NII (Smith, 1994). They do so by specifying proper information processing techniques and protocols that are required to function in the media context constituted by the NII. In essence, these essays outline proper codes of conduct, shape an image of the academic lifestyle of the future, and, perhaps more importantly, encourage a shift from traditional, literate academic practices and methods to electronic ones.

Fourth, the virtual realities constructed through electronic communication systems may significantly alter perceptions, cognitive processes, and intellectual skills. In an electronic culture, the speed with which messages are sent and received, and the use of actual events in the construction of mediated narratives, have the capacity to distort perceptions of physical presence by altering temporal and spatial perceptions (Bolter, 1984; Meyrowitz, 1985; Shapiro & Lang, 1991). When information is stored in a database or broadcast on television, it has no physical place of its own; it is simply "out there," waiting to be accessed by a user or watched by an audience. At the point where information or broadcast image is encountered by the audience, it distorts the perception of physical presence in the immediate situation. So, for example, when the Challenger space shuttle exploded during launching, millions of American viewers who were not actually present at the launch site could nevertheless share in the immediate emotional anguish that followed the fatal explosion.

When the perception of social place is separated from physical place in this manner, many other sociocultural conceptions are also transformed. In this regard, television has demystified distinctions between gender, age, and hierarchical status. For example, the television camera has regularly gained access to physical locations that most viewers are denied access to. In the political arena, this results in a dramatic decrease in privacy. Citizens are privy to nearly all facets of a president's life, both public and private. In this way, audiences are able to identify the natural imperfections that presidents share with other humans. As a result of diminished privacy, television weakens the distinctions of hierarchical status that normally characterize the relationships between group members and, correspondingly, diminishes the assumed authority and power inherent in hierarchically ordered relationships (see, e.g., Meyrowitz, 1985). The same lack of privacy has a similar effect on parent–child relationships, and on relationships between men and women. Access to the staging area, or backstage planning area, demystifies traditional relationships and reveals the strategies and tactics used to enforce the distribution of power (see, e.g., Goffman, 1959). Thus, the virtual realities that characterize communication in an electronic culture transform perceptions of time and space, and as a result significantly alter sociocultural conceptions and relationships that generally organize social groups.

Moreover, researchers have suggested that altered perceptions of time and space may influence cognitive processes and intellectual skills. Donohue and Meyer (1984) have contended that the visual conventions inherent in television, particularly television commercials, "are learned in much the same way language is learned—assimilated through practice, repetition, and by inquiry into what the technique means in the context of what is being viewed" (p. 140). As a result, they have suggested that viewers build expectations that shape perceptions about future experiences with production techniques and styles. These perceptual expectations or patterns encourage certain cognitive processes and intellectual skills while deemphasizing others. Armstrong (1993), for example, has held that background television interferes with two cognitive processes: visuospatial processing and semantic processing. As a result, he has suggested that background television impairs "verbal short-term memory tasks [and] comprehension of difficult or complex written or spoken linguistic material, second language learning, vocabulary acquisition, and rote learning" (p. 68). Similarly, Messaris (1994) has suggested that visual literacy is tied to spatial intelligence. From his perspective, experience with visual media enhances the capacity for abstraction and analogical thinking while deemphasizing analytical thinking. In this way, the virtual realities constructed as technocultural dramas in an electronic culture may indeed be altering cognitive processes and intellectual skills.

These four sociocultural trends are not the only cultural transformations engendered by electronic communication systems. However, they do point out the increasing potential of telecommunication and interactive communication systems to shape perceptions, beliefs, knowledge, and lifestyles. Making sense of the independent and discrete socially constructed realities of electronic culture will, in large part, depend on the ability to make critical distinctions between increasingly complex messages constructed through and transmitted by telecommunication and interactive communication systems.

EVALUATING COMMUNICATION IN AN ELECTRONIC CULTURE

To facilitate the criticism of electronically mediated technocultural dramas, Turkle's (1984, pp. 196–238) chapter, "Hackers, Loving the Machine for Itself," in *The Second Self: Computers and the Human Spirit,** provides a useful critical approach for illustrating how interactive communication systems can be analyzed. We find her essay important, because (1) it implicitly accepts the presence of electricity as a given structural component in the technologies explored, and (2) by concentrating on the inherent power of a technology's structural features to shape messages or order realities, it acknowledges the centrality of rhetoric as an epistemic and ontological force organizing electronic cultures (see, e.g., Lanham, 1993).

Turkle's (1984) chapter explores the culture of the "hackers"—a group of computer specialists who, by nature of their advanced expertise in computer science, may be considered the privileged elite of computer users (p. 199). Much of her analysis is devoted to tracing the sociocultural influences and implications of the computer as a dominant communication system. Turkle has focused her attention on the individual as creative agent, constructing virtual realities from various information sources and technologies. As a result, the central concern of Turkle's analysis is the structural analysis of an interactive communication system. Although the individual as creative agent and information-accessing technologies receive most of Turkle's attention, she has also made some references to audio, visual, and printed modes of information in her structural analysis.

Structural Analysis

Drawing on extensive interviews with hackers at the Massachusetts Institute of Technology (MIT), Turkle (1984) has held that they construct their own way of life through a dyadic relationship between the individual

---

*Our understanding of the critical decisions and processes involved when analyzing electronic communication technologies would be tremendously enhanced, we believe, if Turkle's (1984) volume *The Second Self: Computers and the Human Spirit* is compared to her analysis in *Life on the Screen: Identity in the Age of the Internet* (Turkle, 1995). The comparison, we would suggest, should emphasize changes in her descriptions, interpretations, and evaluations.

and the computer (p. 231). She has argued that hackers, like other people, define themselves "in terms of competence, in terms of what they can control" (p. 208). For the hacker, the computer serves as the medium that supports this human desire for mastery. Indeed, Turkle has suggested that hackers rely on the computer to create microworlds or "cyberspaces" that provide safety, elegance, and controlled fantasy (pp. 207–208, 222). These microworlds provide a social place for the individual hacker and may also sustain collective social action (p. 226). Moreover, because of their expertise with the computer, hackers thrive on the exclusiveness and relative autonomy their computer experiences provide (p. 214).

The microworlds constructed by hackers constitute the technocultural dramas of an interactive communication system, ultimately controlling the hacker's perceptions and activities. Turkle has noted, for example, that hackers delight in increasingly complex computer programs and games. In fact, she has contended that "when systems get complex they become worlds you can live in" (p. 225). Life in a computer-generated microworld is attractive and compelling precisely because the hacker must integrate and manipulate a variety of structural features to understand and control a computer-generated environment. Casting existence in computerized microworlds as a form of gamesmanship, Turkle has suggested that "for the hacker, where what is most central is mastery over complexity, the game takes the form of a labyrinth" (p. 225). Survival in the microworld of the hacker then, depends on their skill at getting through the maze. This skill is tied in large part to the hacker's preference for the recursive, self-referential form of information-accessing technologies, in this case the computer program (pp. 220, 228).

Computer-generated microworlds are inherently interactive (p. 229) and ultimately tied to the operating system created by the hacker (p. 202). Turkle has posited that computerized microworlds are essentially exercises "in composition, structure, and technique" (p. 220). In fact, hackers seem overly interested in the "intricacies of a system" (p. 220). The more intricate the structure of the operating system and the greater the purity of its compositional forms, the more attraction it holds for the hacker (pp. 209, 219). Moreover, Turkle has cast computer operating systems as languages that "mark and protect" cultural boundaries, sustaining the initiated and excluding outsiders (pp. 201–202). In this way, the programming languages developed by hackers drive the computer and simultaneously identify cultures (p. 227).

The individual as creative agent and information-accessing technologies draw a great deal of attention in Turkle's structural analysis. Nevertheless, she has made some slight references to other structural features of an interactive communication system. She has noted, for example, that music plays a central role in hacker culture (pp. 219–221).

However, hackers' tastes in music tend to focus on the structure and technique of composition rather than on its tonal or dramatic qualities. In addition, music is often part of the background of the computerized microworld rather than an explicit structural element used to construct messages in an interactive communication system. Accordingly, music may be cast as a component of the general media context in hacker culture.

Turkle has also made passing reference to the computer screen as the primary receiving and transmitting technology of hacker culture, and thus has indirectly acknowledged the video component of computerized microworlds (p. 233). Further, she has made oblique reference to print as a structural feature of interactive communication systems (p. 225). In sum, Turkle's structural analysis highlights the individual hacker integrating audio, video, and print modes through information-accessing technologies, and illuminates the intertextual quality of interactive communication systems.

Cognitive Analysis

Turkle's (1984) analysis also reflects the conviction that communication systems influence and are influenced by culture. She has, for example, noted that "the structure of computer programming languages . . . encourages different ways of thinking" (p. 227) and has the capacity to alter the hacker's identity (p. 231). As a result, she has concluded that the use of computers constitutes a growing mode of cognitive processing that can shape a culture's understandings, values, and passions (p. 230).

In a complementary fashion, the computer program is most commonly considered a reflection of the programmer's mind. Not surprisingly, Turkle's interview subjects have reported that programming allows them to "build straight from your mind" (p. 235). Moreover, the interaction between hacker and computer supplies the intellectual contact and psychological space that attracts hackers to the medium in the first place (see, e.g., pp. 213, 219). In essence, hackers establish a telepathic relationship (p. 211) with their program that ultimately identifies the complexity of their thinking and defines their cultural boundaries.

Turkle's analysis also deconstructs a number of assumptions about human motivation, human relationships, and reality. She has suggested, for example, that hacker culture deemphasizes traditional rigid hierarchies and academic values by respecting only "the power that someone could exert over the computer" (p. 203; see also p. 205). As a result, Turkle has argued that hackers operate outside traditional systems of social organization (p. 227). Hackers' lives have meaning and purpose because their expertise conveys a degree of uniqueness (p. 216) that sets them apart from others and, quite literally, allows them to "break out of confining situations" (p. 232). Privacy, a social construct designed to order human relationships, is consequently devalued in the hacker's microworld (pp. 232–233).

Turkle has further hypothesized that hacker culture radically transforms human relationships (p. 235). She has noted, for example, that hacker culture overtly denies the necessity of face-to-face interactions (p. 235). In essence, hacker culture may be characterized by "a flight from relationship with people to relationship with a machine" (p. 210). Nowhere is this transformation more apparent than in the antisensuality of hacker culture. Turkle has observed that hackers' relationships with their computers often allow them to replace sensuality with intellectual contact (see, e.g., pp. 215, 219). As a result, hackers display parental concern for their programs, or, as one interviewee put it, their "brainchildren" (p. 235). Under such circumstances, hacker culture provides a safe haven for those whose unique skills might isolate them from the rest of society (p. 216). Hacker culture becomes then, a safe haven, a home where the risks of interpersonal relationships might be minimized (p. 214). In this way, hacker culture allows hackers to safely control their social relationships by encouraging empathy at a distance (see, e.g., p. 217). Turkle has noted, for example, that computer network mailing lists constitute "a community of people who declare themselves to have an interest in common" (p. 223). Accordingly, hacker culture invites its members to share in a virtual reality where human contact can be controlled and rendered predictable.

In sum, the hacker culture deconstructs human motivation and traditional human relationships. The values, customs, and practices of those outside the hacker culture are deemphasized. As a result, hacker culture essentially devalues the real world (p. 213) in favor of the microworld, or technocultural drama, constructed through the interactive capacities of the computer.

Sociological Analysis

Hacker culture exists, however, precisely because it does sustain a common set of social activities and institutions. Turkle (1984) has defined hacker culture as "a culture of mastery, individualism, nonsensuality. It values complexity and risk in relationships with things, and seeks simplicity and safety in relationships with people. It delights in ambiguities in the technological domain—where most nonscientists expect to find things totally straightforward" (p. 223). "On the other hand," Turkle has noted, "hackers try to avoid ambiguity in dealing with people, where the larger culture finds meaning in the half-defined and the merely suggested" (p. 223). This definition clearly indicates shared assumptions about appropriate social relationships in hacker culture and ties those assumptions directly to the structural features of the computer as an interactive communication system. While there is thus common ground for sustaining collective social action, Turkle has specified other features of hacker society.

She has suggested, for example, that hackers share values (p. 216) and engage in initiation rituals that impart appropriate codes for social action to fledgling members of the culture (p. 231). In her discussion of computer system security, Turkle has implicitly identified appropriate protocol for social conduct in the hacker culture and has tied this conduct to the information-accessing technologies and techniques of the system (p. 233).

Turkle's analysis of hacker culture also suggests that it sustains several social institutions with some technological modifications. She has noted, for example, that hacker culture is a culture of loners (p. 213). As a result, she has found that hacker culture is attractive precisely because it sustains a radical individualism where members are free to explore their intellectual capacities within the safety of the computerized microworld (p. 215). In addition, Turkle has cast the hacker as the technological equivalent of the poet or artist, giving free reign to artistic and creative abilities, with the computer substituting for pen or brush (pp. 207, 228). Accordingly, hacker culture can be said to maintain a technological aestheticism. These same aesthetic qualities can also be found in the hacker's fondness for science fiction literature, where the compositional codes reflect the safety, elegance, and controlled fantasy common to the invention of computerized microworlds (p. 222).

Evaluative Analysis

Turkle (1984) provides a window for critically assessing interactive communication systems. Her effort reflects what we take to be an appropriate approach for understanding human communication in an electronic world where active individuals integrate the resources of multiple and diverse information modes into meaningful messages. But Turkle's evaluation of the hacker culture seems foreboding. She has noted that hackers are worried and fearful about their relationships with computers (p. 236). Like all artists, hackers "become inhabited by their medium" (p. 236), and in the hacker culture this intimate involvement seems to deemphasize humanity itself. Turkle has aptly summarized these concerns in her characterization of the outsider's understanding of the hacker's relationship with the computer: "We are surrounded by machines. We depend on them. We are frightened by how powerful they have become. Our nuclear machines have the power to destroy the world. We are suspicious of the new 'psychological machines' and fear the hacker's intimate relationship with his object" (p. 238). As Turkle has concluded, "Its control over him is disturbing because we too feel controlled. We fear his sense of becoming a 'device' because most of us, to one extent or another, have had that feeling. We fear his use of the machine as a safe companion because we, too, can feel its seduction" (p. 238).

| CONCLUSION

In this chapter, telecommunication and interactive communication systems were cast as the dominant communication technologies of an electronic culture. Electricity was identified as the unifying feature of all electronic communication technologies, providing the fundamental resource for active individuals to integrate audio, visual, print, and information-accessing modes. In addition, the sociocultural influences of electronic media were considered. Finally, this chapter examined a critical analysis that isolates some of the possibilities for evaluating human communication in an electronic culture.

Thus far, we have focused our attention on the critical assessment of oral, literate, and electronic cultures and the types of technocultural dramas that each culture might encourage. But we must now ask what lies before us. In electronic cultures, where the receiver shoulders most of the burden of message construction, virtual realities often bear little or no resemblance to the events of everyday life. Indeed, a recent advertisement for *Cybersex*, from a company called Cybertech, invites consumers to purchase a home virtual reality simulator system. The advertisement promises "a sexual encounter that one experiences utilizing the new technology of 'virtual reality,' i.e. not occurring in reality, but with all the sensations, pleasure, and orgasmic response of real sex so faithfully duplicated, as to be virtually indistinguishable from the real thing." What expectations about human communication might we maintain in a future media context where sex partners can be selected and encountered through technological means? What sort of technocultural dramas await us in the emerging communication systems and in communication systems that might be forthcoming? The future of the electronic culture has yet to be determined.

PART IV **A Future Perspective**

We conclude this volume by speculating about the future of communication technologies, and specifically about the future *analysis* of communication technologies. Rather than provide definitive claims, our discussion is intended to be heuristic, to offer a foundation that stimulates by encouraging the development of new ideas, by developing hypotheses and theories, and by fostering new research areas. In Chapter 7, "Analyzing Media Comparatively: Comparative Media Criticism and the Future of Media Criticism," we suggest how particular features of different communication technologies might be compared, and we isolate specific future directions that might emerge as the analysis of communication technologies emerges.

Analyzing Media
Comparatively:
Comparative
Media Criticism
and the Future
of Media Criticism

I n this final chapter, we consider recommendations
for future study. First, we suggest useful ways in
which communication technologies might be
compared. Second, more generally, we suggest future directions for the
analysis of communication technologies.

ANALYZING COMMUNICATION
TECHNOLOGIES COMPARATIVELY

In Part III of this volume, we examined each technocultural drama
discretely and in a separate chapter. This approach allowed us to provide
coherent and detailed considerations of the nature and functions of each
technocultural drama. Another useful approach is to isolate a particular
feature of a communication system and to compare several communication technologies in terms of this single feature or criterion. We have
found such comparisons to be stimulating and to foster new ways of
understanding and thinking about communication technologies as symbolic and cognitive systems.

We want to offer several schemes for making such comparisons. These schemes are offered heuristically. We hope they stimulate new ideas, develop new research areas, and generate new assessments of how communication technologies are and are not affecting the quality of human life and symbol using.

Three of these heuristic schemes are offered in this chapter in Tables 4, 5, and 6.

Table 4, "A Comparison of the Unique Message-Generating Capacities of Major Communication Technologies," draws heavily from Chapters 4–6 of this volume, but we have also taken the liberty of extending several of these comparisons beyond the analyses provided in the earlier chapters. Additionally, we have found this table to be useful in several ways. This table provides, for example, a comprehensive, comparative, and detailed set of examples of the model for analyzing technocultural dramas that we outlined at the end of Chapter 3. Moreover, the table provides vivid illustrations of how communication technologies alter the priorities and emphases of cultural systems. In this regard, a "Structural Analysis" of the oral, literate, and electronic cultures suggests that the nature of the human community has undergone dramatic transformations from an intense, emotional, and physically powerful face-to-face verbal and nonverbal system to the more distant and independent "literate" culture to a human world organized by a series of individually and personally created electronic subcultures. In our view, a comparison of the structural, cognitive, sociological, and evaluative analyses of these technocultural dramas can be equally rewarding and intriguing in terms of understandings and uses of the communication technologies. Beyond these uses of the table, we have also used this "heuristic" moment as a chance to experiment with our own model for analyzing technocultural dramas. Accordingly, we have added a new frame of reference into the model. In Chapter 3, we did *not* consider how a "Postmodern Analysis" of technocultural dramas might be executed. In this more experimental context, it becomes appropriate to examine explicitly how power becomes a major factor affecting the interplay of culture and communication. In this regard, the implicit power system governing the oral, literate, and electronic cultures can be overtly compared and contrasted. In this sense, it becomes tempting to suggest that different kinds of communication technologies might well foster certain forms of government rather than others. In all, then, we have found Table 4 to be provocative and potentially useful in any number of ways.

Table 5, "Symbolic Emphasis and Communication Functions of Media Systems: A Comparison of Communication Technologies," distinguishes communication technologies in terms of rhetoric and information concepts in the discipline of communication. As is true of any academic

TABLE 4. A Comparison of the Unique Message-Generating Capacities of Major Communication Technologies

Messages	Oral culture	Literate culture	Electronic culture
Structural analysis	Sound, rhythm, formula, repetition, nonverbal behavior, and images constitute a cultural system.	Systems of form and formulations of usage constitute culture, with style as a primary canon. Forms are self-motivating.	Space, screen control, and audio manipulations are cast as information units. Electronic systems create isolated subcultural series.
Cognitive analysis	Close relationship to lived human experiences. Emotions and conflicts are "facts" of understanding.	Linear patterns and hierarchies dominate literature. Philosophical interpretations are emphasized over factual reality.	Virtual realities are constituted by technologies and individual choices. A nonsocial process, systems alter consciousness and identity.
Postmodern analysis	Identification occurs through the lived experiences of a hero. Reorganization of society is required to redistribute power.	Grammar as a power system for affecting change.	Uniqueness and safe enclaves are the foundations for social interactions.
Sociological analysis	Community and the family are the primary units of society, with empathetic audience participation as the foundation for all modes of communication.	Intellectual elitism is promoted by the language system.	Telecommunications and interactive systems create new aesthetic institutions and intertextual messages, such as mastery, individuality, and nonsensuality.
Evaluative analysis	Traditions are overemphasized and newly emerging civilized values are promoted.	Literacy should be required for participation in governance systems.	Unstated and unconscious manipulations are processed by electronic technologies that deemphasize notions of humanity.

TABLE 5. Symbolic Emphasis and Communication Functions of Media Systems: A Comparison of Communication Technologies

Criteria	Oral culture	Literate culture	Electronic culture
Data structure	Holistic	Discrete	Concrete referents
Data processing	Transactional	Sequential	Parallel
Causation	Interactional	Linear	Multiple
Figure	Dialectic and irony	Synecdoche	Analogy and metaphor
Quantity of information	High	High (potential overload)	Low (passive learning)
Response set	Social unity	Intense and stimulates thought	Uniform and shapeless
Perception	Present relationships	Specific cases as universals	General orientation and awareness
Logic system	Mimesis or lifestyle identification	Propositional (object claims as true or false)	Appositional or synthesis
Conception of reality	Concrete or naive realism	Abstractions	Selective concrete particulars
View of context	Context-bound	Context-free	Illustrates a specific context
Time frame	Present	Past	Future
Responsibility or guilt attribution	Experiences and circumstances are blamed	Ideas are blamed	People are blamed
Orientation to change	Rejection: burlesque, satire, elegy, and plaint	Transitions: grotesque and didactic	Acceptance: epic, tragedy, comedy, humor, and ode
Degree of participation	Activism	Delayed, critical reactions	Passive learning
Conceptions of situations	Reduces situations to personal needs	Abstracts and classifies situations	Portrayals (visual and auditory) or conceptions of situations
Dramas	Dramas are cast in positivistic terms	Dramas are cast in ultimate terms	Dramas are cast in dialectic terms

discipline, a cluster of different ideas, concepts, and precepts emerge as relevant whenever human communication is discussed. Using Table 5 as our base, we have essentially asked what would happen to the discipline of communication if technocultural dramas were posited as the foundation for examining human communication. If we assume that technocultural dramas are a "given" or a basic foundation of all human interactions, a host of ideas, concepts, and precepts begin to organize themselves. The nature of data systems and data processes, for example, begin to rearrange themselves in some intriguing ways. By way of provoking interest in how the system can be extended, the notions of *holistic data structures, discrete data structures,* and *concrete referent data structures* might be posited as ways of respectively distinguished oral, literate, and electronic communication cultures. Similarly, as we considered how people might cast and organize their ideas or create perspectives when communicating, it seemed to us that certain cultures would find certain rhetorical figures more useful and relevant than others. Hence, the dialectic and ironic posture that Socrates found so important in his face-to-face dialogues with others might be viewed as an essential rather than accidental feature of human exchanges in a predominantly oral culture. Likewise, when people function in a literate culture, we frequently have the sense that a single situation can function as representative of the larger society. Identified formally as the use of *synecdoche* whenever a single event is cast as representing how an entire system functions, the sense of the independent individual itself might function synecdochically whenever people believe they learn "on their own" from their own self-developed and unique reading programs and reactions. And, the content of television series that have no direct meaning on viewers' lives might become relevant and interesting if viewers are able to transform television content—by the use of the figures of analogy and metaphor—when viewing these programs. In all, these distinctions can be understood and used in different ways. Indeed, we hope that the "open-ended" nature of this table provokes, stimulates, and generates new ideas about the nature of human communication processes.

Finally, Table 6, "Modes of Communication as Cognitive Systems" (Chesebro, 1995a, pp. 37–39; 1995b), provides a comparison of more specific communication technologies in terms of the learning requirements as well as the cognitive abilities and understandings generated by each. Although we identify contexts for these comparative assessments, the comparisons are again intended to be as open-ended and flexible as possible in terms of fostering new perspectives and conceptions. Gardner's (1983) concept of multiple intelligence guides this entire analysis. Extending Gardner's scheme beyond what he is likely to recommend, in this table we explicitly suggest that each form of intelligence is directly, if not intimately, linked to specific information-processing systems. Hence, while reading

TABLE 6. Modes of Communication as Cognitive Systems

Mode of communication[a]	Related intelligences[b]	Cognitive requirements and abilities	Types of understandings generated
Speech	Linguistic, bodily-kinesthetic, and interpersonal	Link auditory patterns and referents, identify nonverbal behaviors as meaningful, and coordinate and identify auditory and nonverbal behaviors as social systems.	Understandings are derived from a lived experience; meanings are socially and participatorily constructed; knowledge is the product of an immediate interaction between knower and known; community norms, values and interaction patterns as a definition of what can and should be known.
Reading	Spatial and linguistic	Link discrete visual abstractions and referents, recognize a dynamic relationship between genus and species, process data units sequentially, employ linear inference systems, and process knowledge of the grammatical, rhetorical, and ethical rules governing linguistic constructions.	Understandings are derived by substituting concrete references for the abstractions read; writer and reader are unlikely to share a commonly shared social context for interpreting language; knowledge is created/understood in an individual, unique, and private environment; knowledge takes the form of propositional claims.
Writing	Bodily-kinesthetic, spatial, and linguistic	Construction of linguistic abstractions (categories or genres) from specific, concrete, individual, and lived experiences; visual and physical coordination to generate linguistic units; knowledge of the grammatical, rhetorical, and ethical rules governing linguistic constructions.	Self-understanding; understanding of the relationship between mental concepts and symbolic code—the amount of written elaboration determines how much is learned as well as the degree of retention and comprehension; the relationships among self-understandings, individual adjustment, and social convention; individualism as knowledge.
Music	Musical and linguistic	Link and identify instrumental, lyrical, melodic, chords, and rhythmic acts as patterns; attribute social significance to patterns; and recognize physiological reactions to music as emotionally meaningful social acts.	Formalism; social significance and power of patterns, forms, redundancy, and amplification; problem-recognition and problem-solving through social bonding (i.e., community expression and shared experiences); understanding by maximizing and minimizing emotions; moralizing or ideological posturing (i.e., praise/dispraise) as a revelation process.

(continued)

TABLE 6. *(Continued)*

Mode of com-munication[a]	Related intelligences[b]	Cognitive requirements and abilities	Types of understandings generated
Television	Spatial, linguistic, and musical	Perceive a series of photographic images as continuous motion; understand the conventions employed to create coherence and action within a frame, shot, scene, and sequence; recognize visual figure–ground conventions; and link and identify patterns among visual, linguistic, and musical acts.	Holism; pattern recognition; image recognition; role duplication; figure–ground relationships; multiple causation; metaphoric reasoning; general orientation and awareness (i.e., an inverse relationship exists between television viewing and reasoning: as a symbolic code increasingly approximates real life, the need to make inferences and judgments declines).
Computers	Logical–mathematical and linguistic	Impose one systematic and universal ordering system (analogy) upon all referents, especially language using ones; treat sensorimotor acts as abstractions; and identify/impose universal patterns of interaction within ordering systems.	Reductionism; categorical statements as understanding; deductive argument as understanding; imagery or nonreferential symbolic systems (i.e., isolated, independent, and closed systems) systems as new life and living experiences (e.g., the shift from simulations to virtual realities); decontextualized information.

[a]Every mode of communication occurs within a context, and this context will alter the type of intelligence required to use the mode of communication effectively. Depending on the historical context in which a mode of communication is employed, extremely different skills may be required to derive understandings from the mode. At one point, "books" were individually handwritten scripts that included extensive artwork, in contrast to the contemporary assumption that a book is generated with mass-produced movable type fonts with one alphabet letter per font. Given alternative purposes governing this analysis, at this point in time, these contextual variations are not specified. However, it should be understood that "speech" refers to face-to-face verbal and nonverbal exchanges within a community context; that "reading" and "writing" involve a product produced with mass-produced movable type fonts, and that both reading and writing are relatively private and individual acts with few distractions in the environment; that "television" assumes viewing within the relative privacy of one's home; that "music" refers to a situation in which one is concentrating on the musical stimuli (i.e., music is not a part of the background) by way of stereo or radio, in which noise within the environment is relatively unobtrusive; and that "computers" refers to data-processing activities.

[b]Gardner's (1983) seven forms of intelligence have been employed here to distinguish modes of communication as cognitive systems. Although not linked to communication systems by Gardner, the basic principles of his conception of intelligence are consistent with an effort to link his scheme of intelligence to communication. Gardner has specifically recognized that "central" to his "notion of intelligence is the existence of one or more basic information-processing operations or mechanisms" and that "a human intelligence" can also be "defined" as a mechanism "genetically programmed to be activated or 'triggered' by certain kinds of internally or externally presented information" (p. 64). However, Gardner discusses each type of intelligence discretely. In terms of communication, a single communication system may require that more than one type of intelligence be used to effectively employ a specific mode of communication. To preserve Gardner's understanding that each intelligence is "relatively autonomous" as well as the relative uniqueness of each mode of communication, the intelligences related to or linked with each mode of communication should be conceived as rank-ordered, with the first intelligence understood as far more significant and relevant than subsequent intelligences.

derives its sense of "what is" or knowledge by directly manipulating spatial and linguistic information-processing systems, speech involves a broader range of information-processing systems, particularly in terms of its additional reliance on nonverbal communication, context, and the personal relationship that exists between a speaker and listeners. Thus, Table 6 appears to provide "definitive" claims regarding the relationships between a specific communication technology and various intelligences, but we do hope you will *reconsider* all of these relationships, with an eye to reformulating what is taken to be a given about the understandings and skills required to function effectively with any mode of communication and how each mode of communication also affects what we know and understand.

CONCEIVING A FUTURE FOR THE ANALYSIS OF COMMUNICATION TECHNOLOGIES

The future of media criticism must necessarily emerge from the present. At present, the commitment to and development of communication technologies are far outstripping the investment in describing, interpreting the social consequences of, and evaluating the humane processes and consequences of communication technologies. In many respects, we are creating and using communication technologies before we know how they affect us. Moreover, the *content* or ideas expressed in media continue to attract attention. Communication technologies and media are currently treated as neutral conduits through which ideas are expressed. The assumption here is that ideas can be adequately and comprehensively described, interpreted, and assessed independently of the channels used to convey these ideas. There is little evidence that these ideas significantly and independently alter attitudes, beliefs, or behavior. Nonetheless, attention has been primarily devoted to the degree to which television programs, films, and CDs contain explicit violent and/or sexual themes. The unique message-generating capacity of each communication technology or medium itself has been all but ignored. We find these claims and observations extremely important in terms of what remains to be done about assessing communication technologies as message-generating systems affecting our cultural, symbolic, and cognitive systems. In the balance of this chapter, hypotheses are isolated and detailed that require the immediate attention of those within the discipline of communication.

Pedagogical Implications

Teachers of communication need to reconsider the kind of commitment and the scope of the commitment they have made in terms of communication technologies. Foremost among these decisions have been two decisions that warrant attention: (1) the decision to focus on the content or ideas expressed in any given media system; and (2) the decision to focus on a single mode of communication intrinsically, without adopting a corresponding comparative media or technological base when characterizing a mode of communication. If nothing else, we hope that we have rigorously challenged the reliability, validity, and significance of both a content and a single-medium orientation.

But, more specifically, the nature of public speaking courses needs to be dramatically reconsidered. In terms of the formulations surveyed in this volume, public speaking courses should fall within the domain of the oral culture, with the focus of these courses directed toward teaching students how to function within a context in which verbal and nonverbal dimensions merge speaker, audience, and cultural system into a single, seamless, coordinated, and cohesive social entity. However, as we see these courses, and particularly the textbooks that are designed for these courses, a literate emphasis dominates. Students are given prescriptions for functioning as more effective members of the literate society. Hence, for example, they are provided with detailed instructions for outlining techniques far more appropriate for writing an essay. Likewise, the distinction between oral and written styles has been lost, and questions of style rely profoundly on what works in a formally written essay rather than the kinds of choices that function effectively in oral, face-to-face interactions. Indeed, as some postmodern critics have maintained, literature may now constitute the model for oral conversations. We think it would be beneficial for instructors to reconsider how the distinction between orality and literacy might reflect the lives of their students, and then provide prescriptions for effectively functioning in oral as well as literate cultural systems.

More generally, we hope that a new conception of the effective communicator can emerge. Although any number of criteria should be retained, we would argue for adding another perspective. In our view, a student should be trained in four modes of communication: oral, literate, telecommunications, and interactive. He or she should know the different requirements for each of these modes, should be able to articulate these differences, should be able to function effectively as a communicator in each of these modes, and should be able to know when and how to shift from one mode to another. Students need to be able to function effectively as multimedia communicators.

Multiple Selves in a World of Multiple Communication Technologies

We may confidently maintain that all students should be trained and effective communicators in all of the major communication technologies used to transmit messages in the United States. At the same time, such educational objectives need to be considered in terms of the psychology of the individual and in terms of what a given individual can personally accommodate. In this regard, it is noteworthy that several scholars have maintained that a fragmentation of the individual has been occurring, and that any additional schemes that reinforce this fragmentation will generate permanent counterproductive effects on the individual. There are reasons to believe that use of an increasing number of communication technologies may ultimately have such an effect on the individual.

In *The Saturated Self*, Gergen (1991) has maintained that the meaning of personal life, and therefore one's very self, is changing in profound ways. Gergen has argued that romanticist and modernist conceptions of the self, while still alive, are being overtaken by a new, postmodernist attitude. During the Romantic era, people believed in inner joy, moral feelings, and loyalty. The modern period gave precedence to logic, reason, and observation. The current period of postmodernism, Gergen has suggested, is marked by multiplicity, variety, and change. No longer is the self coherent. Rather, we accept that we have different faces for different roles. Our relationships are played out through impersonal technologies, like faxes or computer bulletin boards, that allow us to reach across the world without looking into the eyes of those we "relate" to daily. It is a condition Gergen has called "multiphrenia" (see, e.g., pp. 73–80, 150, 157, 242).

The increase in the number of communication technologies, the different understandings required to function effectively in each communication technology, and the change in the self required to perform within each medium all suggest that the massive upturn in communication technologies may be contributing to, rather than reducing, the problem of multiple and contradictory personalities for each individual. Indeed, Gergen (1991) has noted that the "phenomenon of *self-multiplication*" has "become paramount in the high-tech era of television" (pp. 54, 55). More generally, he has maintained that the new electronic communication technologies increase the capacity of the individual to be "significantly present in more than one place at a time" (p. 55).

We anticipate that the expanding scope of communication technologies will continue to exaggerate this problem before any clear resolution begins to emerge. For example, cyberspace, with its mix of diverse sensory systems (e.g., sound is used to generate the sensation of touch), can only aggravate the sense of where the self actually is (see, e.g., Rheingold, 1991, pp. 312–377).

At the same time, rather than reduce the sophistication and power of potential communication technologies, we anticipate that psychological mechanisms will be considered and adapted to deal with the question of multiple and contradictory self-images. Indeed, more attention is already being devoted to questions such as self-reflexivity, irony, and play as critical constructs to explain experiences that individuals encounter. As the complexity of communication technologies increase, we expect a corresponding increase in psychological constructs used to explain the sense of the self in these mediated communication systems. Ultimately, however, we would hope for a reconciliation between the culture (and its suspicion of criticism as self-serving) and criticism (as a valuable way of understanding both culture and the self). Specifically, thoughtful and informed critical views and statements should and can function as a way of creating a coherent and integrated cultural system for the individual.

Revisiting the Meanings, Roles, and Functions of Criticism, and Media Criticism in Particular

In Chapter 3, we examined the roles and functions of criticism in general and media criticism in particular. As we concluded our analysis of the fragmented self in a postmodern and multicommunication technology environment, we also noted that criticism, and media criticism specifically, might play an especially significant role in promoting a sense of integration and coherence for each individual. In our view, all of these formulations are instructive, and they are indicative of the largely untapped potential that critical analyses can perform in the everyday life of most people.

At the same time, criticism now "competes" with several other knowledge systems, such as science and ethnography. Within these scientific and ethnographic frameworks, it may be particularly useful to reconsider the most fundamental nature of criticism with an eye to reinvigorating the meanings, roles, and functions of criticism and to reestablishing the power and significance of criticism in daily life.

Toward this end, it is necessary to suggest how criticism can be viewed and will ultimately function as a knowledge system that is equivalent to science and ethnography. The quest for such a formulation is, of course, a long-standing and now a classic concern. At this point, then, we can only suggest a perspective that might be explored as media criticism evolves and develops over the next several decades.

In our view, science, ethnography, and criticism each focus on a different object of study.

Science is committed to the study of the world of phenomena, and from a human perspective, science and social scientists predominantly examine *behavior*. For the scientist and social scientist, a human being's manner of conduct or stimulus–response patterns provide insightful clues into the nature of human motivation, human institutions, and future human action.

Ethnographies focus on the study of human environments and lifestyles. For communication ethnographers, the object of study is the *symbol using of human beings in everyday communication*. Accordingly, ethnographers are more likely to investigate particular intentions rather than intentionality as an epistemic issue; the settings or situations in which communicators assume that a single reality exists independent of perception; imperative actions rather than dialective exchanges; the full range of verbal and nonverbal behaviors that occur continuously and prereflexively in face-to-face interactions; and interactions in which communicators believe that a correspondence of meanings exists among participants and which functions regularly within specific geographic environments and social relationships. In all, the communication ethnographer seeks to identify the pattern of symbol using characterizing everyday communication.

Given the social scientist's focus on behavior and the ethnographer's attention to lifestyle, the critic's role may seem duplicative and repetitive. However, in our view, the critic uniquely investigates the *values* that humans employ. In this regard, critics examine the process that leads humans to appraise, rate, and scale the utility, usefulness, importance, and general worth of themselves, others, environments, and all items or phenomena within their environments. The study of this valuing process can lead to considerations of the extremely wide range of evaluation systems used by human beings and the rank-ordering or priority system humans employ when they are evaluating, as well as the interpersonal, group, subcultural, national, and cross-cultural differences in the evolution, use, and significance of value systems.

Cast as the knowledge system responsible for the examination of human valuing, criticism would fulfill a powerful role in the study of human symbol using, for symbol using itself is uniquely defined as the way in which human beings incorporate evaluations and assessments into their signal using. As media systems continue to grow in size and use, the study of valuing in the content and formatting systems of contemporary communication technologies would seem to be one of the most significant projects that a researcher might undertake.

THE INTEGRATION OF COMMUNICATION TECHNOLOGIES

Beyond the importance of human use of communication technologies, it is also useful to ask if trends are discernible in terms of media systems themselves: *Are communication technologies increasingly specializing? Are communication technologies merging and converging? Are communication technologies increasingly or decreasingly operating within discrete social contexts?* There are, of course, no easy answers to these tremendously important, but complex, questions.

However, some have found coherence, even stages of development and coherence, among the emergence of the diverse communication technologies we have discussed. For example, Braman (1993) has isolated three stages in the development of the "information society." As he has explained,

> The information society, however, did not emerge full-blown. Rather, three developmental stages can be identified. A first, beginning in the middle of the 19th century, was characterized by the *electrification* of communication. The second, beginning in the late middle 20th century, was characterized by the *convergence* of technologies and by *awareness* of the centrality of information to society. The third stage, beginning with the 1990s, is characterized by *harmonization* of information systems with each other, with systems across national borders, and with other social systems. (p. 133)

Technologically, we have already surveyed a great deal of evidence that suggests that these three stages are discrete and significant. At the same time, we remain somewhat skeptical that the pedagogical and individual issues discussed at the outset of this chapter, and the disciplinary issues that close this chapter, will be as easily resolved as some have implied. Nonetheless, it is difficult to dispute the claim that the technologies themselves can be increasingly integrated.

Reconsidering the Meaning of Media Studies as an Area of the Discipline of Communication

As we see it, dramatic changes are occurring in the scope and impact of communication technologies in the United States. Correspondingly, the discipline of communication needs to devote extensive discussion to how media studies should be defined, focusing on what should be examined and how research and critical analyses of these objects should be undertaken. In this regard, we hope that the considerations shaping the May 1995 issue of *Communication Theory* are only the initial step in a long series of discussions regarding the nature of media studies. This issue reviewed a host of concerns that deserve the attention of the entire discipline of communication. For example, Schoening and Anderson (1995) outline "six premises and the implication for media analysis":

1. "World" and its Meanings (i.e., "Reality") Must Be Both Produced and Maintained in Consciousness. Correspondingly, the Meaning of Media Content Has No Autonomous Existence and Must be Brought Into Being in Deliberate Ways. . . .
2. The Interpretation of Media Content Emerges Through the Per-

formance of Identifiable Lines of Action That Signify "What is Being Done." As Such, Media and Content, as With Any Material Fact, Will Always Be Made "Real" in Conjunction With Some Known and Knowable Activity. . . .

3. The "Decoding Activities" Produced by Signifying Lines of Action Shift as Those Lines are Reinterpreted and Performed in Different Routines (In the Same Way that a Sentence Changes its Interpretive Power by its Position Within a Paragraph). As Such, the Meaning Potential for Both Content and Interpretive Practices Remains Open-Ended. . . .

4. "Knowledge" is Made Concrete in Signifying Practices and is, thus, Necessarily Relative To, and Contingent Upon, the Here-and-Now Performance Characteristics of a Given Social Collective's Routine. The Domain of Media Effects is a Cultural Production, but an Effect is Known Only in Local Settings. . . .

5. All Lines of Action—Including Media-Related Behaviors—Are Dialogic and Improvisationally Enacted by Local Agents as a Partial Expression of a Collectively Held Semiotic of Action. . . .

6. The Social Scientist Has an Ethical Obligation to Contribute Understanding About Social Practices to the Social Worlds Investigated. The Social Action Media Researcher, Therefore, is Obligated to Inform Social Groups About the Potentials and Consequences of their Media Practices. (pp. 99–111)

There is no question that these six propositions are potentially controversial. We may not want, for example, to retain the focus solely on material practices, as recommended in this analysis. Yet there seems to be little question that these six propositions argue effectively for the study of media technologies and content, the cultural systems that support media systems, and how local routines and texts fashion and adapt media systems, as well as how the social background of media researchers and media critics affect their analyses. Beyond these issues, the analyses of Streeter (1995), Schoole (1995), McLaughlin (1995), and Kellner (1995) outline a series of claims and issues that should affect the agenda of members of the discipline of communication.

| CONCLUSION

As they continue to increasingly influence virtually every human being in more ways every year, communication technologies invite responses, particularly critical evaluations of the symbols and cognitive systems human beings are to live with, by, and through on a daily basis. This volume has been designed to stimulate these evaluative processes.

Glossary

This glossary identifies and describes a set of basic key terms that we use in earlier chapters to explore communication technologies. To deal more effectively with the full scope and meaning of the transformations affecting communication technologies, the five key terms persistently used throughout this book are examined here. These five key terms are *communication, cognition, culture, technologies*, and *technologies as communication systems*.

We examine each of these key terms in two ways. First, we use each of these terms as an opportunity to define human processes that are profoundly ambiguous and complex. A cluster of inconsistent, if not contradictory and ambiguous, meanings are associated with these key terms. Second, we also use each key term as an opportunity to reveal explicitly the *perspective* we use when explaining how we view a subject or the interrelationships among the parts of a subject.

We believe that definitions and perspectives are intimately related. The description we offer here of each of these key terms provides a definition of a complex human process and system. We seek definitions that distinguish and demarcate a subject from other phenomena, specify the essential qualities and meanings of a subject, and provide a reliable and valid conception of a subject. However, most subjects can be understood in several ways or from several different perspectives. Accordingly, in one sense, a definition can merely indicate how one personally looks at and thinks about a subject.

| COMMUNICATION

Unfortunately, no single definition for the word *communication* exists. Dance and Larson (1976, pp. 171–192) compiled 126 definitions of *communication* from "diverse fields and diverse publications," each of which had been "subjected to some prior scrutiny and expert evaluation" (p. 171). Each of these definitions highlights a different but important feature of human communication.

Some definitions highlight the intentional nature of communication: " 'Communication' implies that the sender's visual signal is intentional and the receiver's interpretation assumes that intentionality" (p. 177). Some feature communication as a process: "Communication is a word that describes the process of transferring meaning from one individual to another" (p. 174). Some stress the content of communication: "Communication is the transmission and interchange of facts, ideas, feelings, courses of action" (p. 173). Some underscore the interactional nature of communication: "Communication must be two-way, for the response is part of the process" (p. 172). And some emphasize the outcomes and consequences of communication: "Communication [is the] study of the ways by which men affect each other and the interactions of those systems of influence" (p. 178).

It is often useful to characterize communication as an intentional process of transmitting certain facts, ideas, feelings, and courses of action to others as a way of interacting with them and ultimately affecting their attitudes, beliefs, and actions. Indeed, conceived in this fashion, any of these dimensions might be highlighted to characterize how and to explain what occurs when human beings communicate. A definition ultimately is selected because it is useful in terms of specific objectives (see, e.g., Chesebro, 1985a).

Because we are concerned with the media and how they function as communication systems, we are drawn to a definition that is flexible enough to account for human responses to media and technological systems but also allows a critic to examine how and why they might become message-generating systems that affect their users. Accordingly, we frequently have found it useful to think of communication as rhetoric, and to define rhetoric as "the human effort to induce cooperation through the use of symbols" (Brock, Scott, & Chesebro, 1990, p. 14).

Yet even a definition as broad as this is troublesome because it implies that all rhetoric is directly and immediately the product of human activity. When dealing with media systems, such a view can be unduly restrictive because the definition suggests that critics should focus their attention primarily on human efforts. However, a critic needs to be open to the possibility that media and technological systems can generate messages that humans do not expect. Williams (1982) has argued, for example, that computers are capable of "*acting upon*" human messages, altering and re-forming them, and providing useful feedback to human beings because they have altered the human messages in unexpected ways (p. 108). Similarly, Chesebro and Bonsall (1989, pp. 179–210) have argued that classes of "semi-intelligent" and "intelligent" computer programs exist, because these programs can acquire sensory data, store and accumulate information, process conclusions, create new links among existing patterns of information, function efficiently, and complete a wide range of activities simultaneously.

A definition of communication ultimately is required, then, that alerts the media critic to a full range of message-generating possibilities, whether the messages are of human or technological origin. With this goal in mind, we have found it useful to define communication as any symbol-generating activity. In this sense, we are interested in communicative exchanges that go beyond the use of signals. A signal is a concept (verbal and/or nonverbal) that specifies or stands for the physical existence, physical characteristics, and/or physical functions of a referent or entity. Going beyond mere signal using, we focus our attention on symbol using. A symbol is a verbal and/or nonverbal concept or construct that specifies, emphasizes, highlights, or reveals the values associated with a referent (see, e.g., Blankenship, 1966, 104–108; 1968, pp. 12–19). In other words, to focus on symbol using is to focus on the value judgments or connotations conveyed by words or nonverbal action (see, e.g., Burke, 1961/1970, pp. 9–10).

Besides serving as a definition, it is also useful to note that this conception of communication can also constitute a perspective or view from which human activities and products are explored and studied. Because we cannot examine human activities and products from all orientations simultaneously, we must necessarily examine them from a single viewpoint. Given our commitment to the study of communication, as we have defined it, this commitment also reflects how we approach the study of human activities and products. For us, the meanings of these human activities and products are determined by focusing on symbol using, that is, the process whereby human activities and products are defined and operationalized as verbal and nonverbal actions that convey value judgments and connotations to others. This definition also constitutes the perspective or foundation for how we construct our understanding of human activities and products.

Yet the decision to focus on symbol using has raised several critical issues. One of the key issues is what the sources of symbols are today. Specifically, we have sought to determine whether or not technologies function as message-generating systems, and we have been particularly interested in technologies conceived and designed to be *communication technologies*. This brings us to a description of cognition.

| COGNITION

We are interested in the question, "What are the effects of communication, media, and communication technologies?" A host of perspectives are possible when answering this question. Our perspective is predominantly cognitive. We are interested in how communication technologies affect shared social understandings. Two concerns deserve attention here.

Our first concern is how communication technologies create a sense of "what is" or influence what people believe "exist as real." From our perspective, a communication technology affects what and how people think—that is, they influence what is perceived, apprehended, and understood. This is not to say that communication technologies are the only factor affecting perception, apprehension, and understanding, but they are a critical factor that exerts independent influence on human cognition. In sustaining this claim, we can consider diverse research findings, social scientific and cross-cultural ethnographic reports, and critical analyses of symbol-using patterns. This perspective is illustrated in Carpenter's (1960) comparison of media systems. He has initially maintained that, "English is a mass medium. All languages are mass media. The new mass media—film, radio, TV—are new languages, their grammars as yet unknown. Each codifies reality differently; each conceals a unique metaphysics" (p. 162). In a more extended statement, Carpenter has concluded,

> Each medium, if its bias is properly exploited, reveals and communicates a unique aspect of reality, of truth. Each offers a different perspective, a way of seeing an otherwise hidden dimension of reality. It's not a question of one reality being true, the others distortions. One allows us to see from here, another from there, a third from still another perspective; taken together they give us a more complete whole, a greater truth. New essentials are brought to the fore, including those made invisible by the "blinders" of old languages. (pp. 173–174)

Our second concern is what people understand collectively. Unique individual reactions are important, but our focus is on how "audiences" or social units come to know or to understand, specifically, the following issues:

- How social units come to understand themselves and their identity as social units (e.g., social constructions of reality);
- How they label objects, ideas, feelings, and events (e.g., symbolic constructions of reality);
- How these labels selected determine social consequences, such as social bonds and social divisions;
- How labeling its social consequences creates values, and how these values begin to define and to determine what people know and understand as "real"; and
- How communication technologies have changed and continue to change historical and cultural developments of entire groups and societies of people.

This cognitive approach is strongly influenced by Gardner (especially 1983), but others have played a critical role in how we identify and deal with cognition (see, e.g., Armstrong, 1993; Burleson & Waltman, 1988; Desmond, 1987; Elliott, 1979; Elliott-Faust & Pressley, 1986; Gorman & Carlson, 1990; Greene, 1988; Greenfield & Beagles-Roos, 1988; Harris, 1994; Olson & Bruner, 1974; Staats, 1968; Wilson, 1974).

In all, we suggest that every communication technology is a channel or medium used as a perspective for describing, interpreting, and evaluating human communication. As a "channel," it ultimately constitutes a symbolic orientation because large-scale, complex sociotechnical systems realign and reprioritize cultural values. In this regard, communication technologies organize society because they are time- and generation-bound, exerting their influence until subsequent communication technologies emerge to replace them. In this sense, large-scale, complex sociotechnical communication systems affect, if not regulate, cultural systems, especially in terms of the cognitive and aesthetic schemes they generate. Accordingly, communication technologies engender their own critical systems for assessing phenomena and experiences.

| CULTURE

Little agreement exists regarding the meaning of the word *culture*. Some 45 years ago, in 1952, Kroeber and Kluckhohn identified 164 definitions of *culture*. They concluded, "Each individual selects from and to greater or less degree systematizes what he experiences of the total culture in the course of his formal and informal education throughout life" (p. 157). Since 1952, the number of identifications has increased geometrically.

Although culture remains an ambiguous concept, Goodenough (1971) has provided some preliminary observations that narrow the number of characteristics associated with it. He has argued that the study of culture directly involves an examination of those social structures that reflect, reveal, and emphasize the standards common to members of a societal system. The content of a cultural system is thus composed of the ways in which people have organized their experiences of the world, the system of cause and effect relationships, the hierarchies of preferences, and the recurring purposes used for achieving desired futures. As Goodenough summarized his view, "Culture, then, consists of standards for deciding what is, standards for deciding what can be, standards for deciding how one feels about it, standards for deciding what to do about it, and standards for deciding how to go about doing it" (p. 22). Accordingly, for Goodenough, culture becomes the study of a set or pattern of societally shared standards.

From a communications or symbolic orientation, four implications regarding cultures are appropriately highlighted.

First, communication technologies are one of the key "webs" creating a cultural system. *Webster's New Collegiate Dictionary* (1981) has defined a culture as an "integrated pattern" of human behavior (p. 274). The "integrated pattern" or "web" that makes a culture is constituted by a number of factors. In our view, communication technologies are one of the social institutions that form this integrating pattern or web. In other words, the pattern of persistently employed codes, techniques, and channels through which people interact exerts an influence on the culture or the values that one generation passes on to the next generation. MacKenzie and Wajcman (1990) have argued that technology cannot be construed as "independent" of society. At the same time, they have maintained that technology is one of a variety of factors that cause cultural change.

Second, communication technologies separate generations. Gumpert and Cathcart (1985) have maintained that "differing world perspectives and human relationships are as much a matter of media gaps as they are generational gaps. We believe that all of us are separated from our past and connected with our future more by media experience and exposure than by chronological years" (p. 23). In this view, people are "born into" an era characterized by certain media systems. Hence, people develop a "media consciousness" from existing media systems, learning the "media grammar" or "rules and conventions based upon the properties which constitute the media" they use, ultimately acquiring "media literacy" or "ability to meaningfully process mediated data" in the media systems with which they are familiar (Gumpert & Cathcart, 1985, p. 23). According to Gumpert and Cathcart, a specific media consciousness leads to a particular worldview or way of understanding experiences:

> It is our position that: 1) there is a set of codes and conventions integral to each medium; 2) such codes and conventions constitute part of our media consciousness; 3) the information processing made possible through these various grammars influence our perception and values; and 4) the order of acquisition of media literacy will produce a particular world perspective which relates and separates persons accordingly. We will examine the nature of media grammar and acquisition of media literacy, and then discuss how this produces media gaps or a "weltanschauung" peculiar to each group's media consciousness. (p. 23)

Third, each communication technology generates distinctive standards and methods of criticism. Every new technology questions the existing culture because it offers alternative ways of doing things, and in the process, it highlights different features of the environment, generates a new vocabulary for talking about experiences, and ultimately leads to alternative explanations of experience as well as raises questions about existing aesthetic and ethical systems. Batra (1990) has argued,

> Technology, like art and literature, is criticism of a society because it, too, questions the assumptions on which society is based by suggesting alternatives. It tends to break down orthodoxy and ideological rigidities by making the impossible plausible and by freeing the human's imagination. It is not surprising, therefore, that scientific and technological imagination has supplanted the artistic imagination as we enter the last decade of the twentieth century. (p. 20)

Batra's conclusion is particularly noteworthy: "It is in this sense that technology is not merely technique or applied science but a critique of culture" (p. 20).

Fourth, types or kinds of "media cultures" exist. In 1983, Snow argued for the recognition and existence of what he identified as a "media culture." In his view, "a culture" is "constructed and altered continuously through the linguistic and interpretative strategies of media . . . [that] requires understanding the language of media and what perspectives are employed in selecting and interpreting phenomena" (p. 7). We use this notion as a way of identifying and characterizing four media cultures that have dominated the history of human communication. We maintain that four of these media cultures have dominated human experience, and we identify these media cultures as the *oral culture, literate culture, telecommunications culture*, and *interactive culture*. These four media cultures are the objects of study in this volume, but we recognize that the telecommunications and interactive cultural systems are in the process of merging, and we believe they are likely to function as one in the near future. Accordingly, depending on the objective at hand, we have sometimes found it useful to treat the telecommunication and interactive cultures as discrete and independent communication systems, while at other times, it is more informative to recognize the emerging relationships between them (in that case, we use the term "electronic culture" to link them).

A cultural system controls many human processes. In part, it establishes what we learn, how we learn, when we learn, and even where it is appropriate to learn. In this regard, culture is intimately linked to human cognition. We are intrigued by this culture–cognition relationship. We are equally convinced that communication systems create, sustain, and alter culture–cognition relationships. Our tendency is to examine communication technologies as cultural systems that affect cognition. In this sense, communication, cognition, and culture all become our key and interrelated critical terms. However, these relationships ultimately remain incomplete without a consideration of how technology affects them.

| TECHNOLOGY

Technologies have become increasingly complex. We associate five interrelated dimensions with the meaning of *technology*.

First, a technology is instrumentation that extends human activity. In the mid-1970s, Olson (1974) formally defined a technology as "any tool or artifice that amplifies or extends man's muscular or intellectual abilities" (p. 12). In this context, McLuhan (1964) implied that media systems were technologies because each medium is an "extension" of our human senses (e.g., television is an extension of human sight). McLuhan also reasoned that insofar as it extends human sensory systems, a media system "shapes and controls the scale and form of human association and action" (p. 23). McLuhan likewise observed that technologies that extend the human senses might easily go unnoticed because a "medium blinds us to the character of the medium" (p. 24). In all, to be classified as a technology, an instrument must be able not only to mimic, but also to extend human activity.

Second, a technology generates information. Porat (1977) wrote: *"Information is data that have been organized and communicated. The information activity* includes all the resources consumed in producing, processing and distributing information goods and services" (Vol. 1, p. 2).

Third, as they generate information, technological systems become epistemic. In other words, technologies do more than generate mindless "data," they also construct social knowledge. In this regard, Rogers (1986) has provided a basic definition: *"Communication technology* is the hardware equipment, organizational structures, and social values by which individuals collect, process, and exchange information with other individuals. Certain communication technologies go back to the beginning of human history, such as the invention of spoken language and such written forms as the pictographs on the walls of caves" (p. 2). However, Rogers has also maintained that communication technologies are knowledge systems. Underscoring the cognitive role of technologies in linking human beings to the world, Rogers (1986) has argued,

> All communication technology extends the human senses of touching, smelling, tasting, and (especially) hearing and seeing. Such extensions allow an individual to reach out in space and time, and thus obtain information that would not otherwise be available. Media technologies provide us with 'a window to the world,' and as a result we know more about distant events than we could ever experience directly. (p. 2)

Fourth, in addition to generating social knowledge or epistemic understandings, technologies also generate ontologies, or determine what human understandings of the nature of existence will be. Questions of existence are frequently resolved by appealing to everyday experience. These appeals ultimately assume a sense of what exists in the applied domain or what is assumed in everyday communication (Chesebro & Klenk, 1981, pp. 327–328). In this regard, *Webster's Third New International Dictionary of the English Language Unabridged and Seven Language Dictionary* (1986) associated technology with applications in its definitions of technology: "technology: (1) the terminology of a particular subject: technical language; (2a) the science of the application of knowledge to practical purposes: applied science; (2b) the application of scientific knowledge to practical purposes in a particular field of study; and (3) the totality of the means employed by a people to provide itself with the objects of material change" (Vol. 3, p. 2348).

Fifth, and last, a technology can possess a self-generated sense of purpose. Perhaps it is consoling to believe that technologies vary in the degree to which they are under their own control. Yet in certain cases—but not all—technologies emerge because they can formulate their own objectives. Such technologies include artificial intelligence systems such as heuristic, planning, backward-chaining, concept-learning, and parallel programs (Chesebro & Bonsall, 1989, pp. 179–210). Similarly, in the case of communication technologies, machines are designed to generate information when circumstances rather than people warrant it because independent action by machines is just what makes communication technologies useful. Williams (1982) has explained that we often think of computers as "giant calculation devices," but he has argued that they can also "qualify as a *communications technology*" (p. 108). As Williams has explained, "They are capable of taking our messages and giving them back to us or others, as does any communications device. But unlike any other communications device, they are capable of acting upon them in a manner defined by an extension of our own human intelligence" (p. 108). Williams has concluded: "While other communications technologies extended the range of our human messages, the computer allows us to extend our human capability for acting upon messages" (p. 108).

In this regard, it is helpful to define a technology as an entity that can become independent of the original human purpose for creating it. In her discussion of "technology as a form of consciousness," Miller (1978) has argued that media systems can be tools or technologies. Tools, Miller has noted, "extend the immediate biological capabilities" of human beings, and when media systems function as tools, they are appropriately understood as extensions of the self (p. 229). However, Miller has argued, media systems can also be technologies. As technologies, media lose their original purpose and begin to function independent of humans. When media systems function as technologies, by definition, it is no longer possible to characterize them solely in terms of the purpose attributed to these systems by human beings (p. 229).

In all, a technology is a complex system, defined by five interrelated dimensions that ultimately create a paradoxical environment for human beings. On one hand, a technology is a human product. Technologies are instrumentations that extend human activity and generate information as well as understandings that are useful to the human being. On the other hand, a technology can realize objectives independent of their original human purposes as well as provide knowledge and understanding humans have not created but nonetheless find useful.

Yet beyond identifying the essential features of a technology definitionally, these five interrelated sets of characteristics also reveal a perspective that can be used to examine human activities and products. The focal point for exploring human activities and products can begin with an exploration of the instruments human beings use, the information and understandings these instruments generate, and how human instruments can become means and ends unto themselves, gaining independent intentional and knowledge-generating capacities. Accordingly, as a perspective for studying human activities and products, technologies can function as the direct and immediate objects of study from which an exploration of the human condition begins.

Here we have considered how communication and technology are explicitly related; next, we determine how these two concepts interact or function together.

TECHNOLOGIES AS COMMUNICATION SYSTEMS

How do communication and technology fit together conceptually and in practice? The concept of a *communication technology* is a bridge or link between them. The concept has been useful, but the meaning attributed to it has varied.

In a particularly narrow sense, a *technology* can be understood as "the tangible products of science" ("Technology," 1987, Vol. 11, p. 601). Ample illustrations of this conception are provided in *The Timetable of Technology: A Record of the 20th Century's Amazing Achievements* (Ayensu, 1982). Ayensu has described 457 communication technologies invented between 1900 and 1982, and he has predicted the development of another 27 communication technologies between 1983 and the year 2000. Emphasizing specific outcomes and mechanisms of scientific innovations, these technologies ranged from the first time a telephone link was used to transmit human speech on radio waves in 1900, to the development of a random access computer system by Alan Turing in 1945, to the introduction in 1982 of Kodak's 15-exposure high definition emulsion disc, with coded information to facilitate processing and printing.

In a broader sense, a *technology* has been understood as "the attitudes, processes, artifacts, and consequences associated with scientific research" (*Technology*, 1987, Vol. 11, p. 601). A similar conception of technology exists in the vernacular as "the totality of the means employed by a people to provide itself with the objects of material culture" (*Webster's Third New International Dictionary*, 1986, Vol. 3, p. 2348) or as "the totality of the means employed to provide objects necessary for human sustenance and comfort" (*Webster's New Collegiate Dictionary*, 1981, p. 1188). In this view, reflected in *The Timetables of History: A Horizontal Linkage of People and Events* (Grun, 1982), every technology must be defined and understood in terms of a particular history, political era, literature, theater, religion, philosophy, learning system, system of visual arts, music, and pattern of daily life.

This broader understanding of technology has influenced how several media scholars have defined "communication technology." Anderson and Meyer (1988) have defined a communication technology as "the knowledge, practices, devices, and texts that constitute the understanding of a domain of human action as seen from the perspective of the device itself" (p. 339). They have noted that a communication technology "is the sum of the tools and knowledge that generate an accepted way of doing things" (p. 68).

This broader view of communication technology guides our approach to media criticism in this volume. Indeed, we hold that media criticism is the analysis of communication technologies, and we see any communication technology as a complex of three interrelated dimensions.

First, a communication technology is a channel. In his classic volume, *The Process of Communication*, Berlo (1960) defines a channel by raising four "channel questions":

1. What kinds of messages should be transmitted orally in the classroom?
2. What kinds of messages should be transmitted visually, through books?
3. What kinds of messages should be transmitted visually, but nonverbally, through pictures, rather than words?
4. What kinds of messages should be transmitted physically, through touch, having students actually perform certain tasks, examine and manipulate certain objects, etc.? (p. 67)

All of these "channel questions" deal with the mode of communication rather than the ideas expressed. When a technology is viewed as a channel of communication, the technology is no longer just a component or variable in the communication process. As a channel in the communication process, a technology is a perspective for examining the entire communication process.

Second, a communication technology is a type of symbolic action. Physically, technology is the "process of applying *power* by some *technique* through the medium of some *tool* or *machine* to alter some material in a useful way" (Roland, 1992, p. 83). Yet this technological process is also profoundly symbolic because its verbal and nonverbal constructs attribute values to or convey connotations which create a shared universe of subjective associations related to phenomena. Thus, the technological process functions on two levels. On the most immediate phenomenal and referential level, a technology changes the power relationships, the practical and mechanical processes, and the material outcomes in and the status of a human environment. On a more abstract and symbolic level, technological processes also create realignments in the human environment, generate new vocabularies for describing these changes, and suggest new priorities or values among how, when, and why certain actions should be taken. Noting that "Technologies are forms of life," Winner (1980) has maintained that "as regards large-scale, complex sociotechnical systems, we do not use technologies so much as live them" (p. 125). Misa (1992) has reported that "through our technologies we construct ourselves" (p. 6).

Third, every communication technology generates an aesthetic unique and appropriate to it. To gain acceptance, a new communication technology initially may imitate an older, more familiar medium, but ultimately each comes into its own only when the new communication technology generates its own forms. As Batra (1990) has aptly observed: "Technology enables humans to create new art forms and demands new aesthetic standards and new approaches to criticism as one can see in television programming" (p. 17).

In this view, a communication technology is a channel that links human beings together. These links create new forms of symbolic action and new senses of what is and is not aesthetic. As a perspective or way of studying human activities and products, these symbolic and aesthetic links can constitute the point of departure used to understand the human situation. We have found it immensely rewarding and intriguing to use communication technologies as our point of departure for examining human activities and products.

| CONCLUSION

In this glossary, we described five key terms: *communication, cognition, culture, technology,* and *technologies as communication systems.* The descriptions suggested how these concepts can function in mutual and interrelated ways when we undertake critical analyses of communication technologies as symbolic, cultural, and cognitive experiences. Overall, we defined *communication* as the human effort to induce cooperation through the use of symbols, but noted that if the communicative efforts of technologies are to be accounted for, then communication must be understood as a symbol-generating, cultural, and cognitive experience.

The quest to define *communication, technology,* and *communication technology* has also generated certain persuasive ends. An understanding of the full scope and pervasiveness of these communication technologies requires that each of us has little choice but to describe, interpret, and evaluate these communication technologies, if we are to retain control over and improve the quality of our own lives. Attention cannot be directed solely to the content of the diverse communication systems we encounter; we must also direct our attention to the channels and mechanisms from which these messages emerge. Communication channels are not neutral conduits of information; they are highly selective information gathering systems that shape and configure information as necessitated by the production and processing requirements of each communication system. Accordingly, communication technologies themselves should be viewed as defining and determining what each of us understands and knows as individuals and as members of society. If we are to function effectively within these systems, each of us will have little choice but to function as a media critic of these "technocultural dramas."

References

Aden, Roger C. (1994, Summer). "Back to the Garden: Therapeutic Place Metaphor in Field of Dreams," *Southern Communication Journal, 59,* pp. 307–317.

Affron, Charles. (1982). *Cinema and Sentiment.* Chicago: University of Chicago Press.

Alter, Robert. (1985, September). "The Poetry of the Bible." *New Republic,* pp. 28–31.

Altheide, David L. (1991). "The Impact of Television News Formats on Social Policy," *Journal of Broadcasting and Electronic Media, 35,* pp. 3–21.

Altheide, David L., and Robert P. Snow. (1979). *Media Logic.* Beverly Hills, CA: Sage.

Anderson, James A., and Timothy P. Meyer. (1988). *Mediated Communication: A Social Action Perspective.* Newbury Park, CA: Sage.

Andrews, James R. (1968). "The Rhetoric of History: The Constitutional Convention," *Today's Speech, 16,* pp. 23–26.

Andrews, James R. (1983). *The Practice of Rhetorical Criticism.* New York: Macmillan.

Armstrong, Edward D. (1993). "The Rhetoric of Violence in Rap Music," *Sociological Inquiry, 63,* pp. 64–83.

Armstrong, G. Blake. (1993). "Cognitive Interference from Background Television: Structural Effects on Verbal and Spatial Processing," *Communication Studies, 44,* pp. 56–70.

Arnheim, Rudolf. (1974). *Art and Visual Perception* (Rev. ed.). Berkeley: University of California Press.

Aronowitz, Stanley. (1988). "Postmodernism and Politics." In Andrew Ross, Ed., *Universal Abandon? The Politics of Postmodernism* (pp. 46–62). Minneapolis: University of Minnesota Press.

Asimov, Isaac. (1963). *The Human Brain: Its Capacities and Functions.* New York: New American Library.

Auer, J. Jeffrey, Ed. (1963). *Antislavery and Disunion, 1858–1861: Studies in the Rhetoric of Compromise and Conflict.* New York: Harper & Row.

Auerbach, Erich. (1953). *Mimesis: The Representation of Reality in Western Literature.* Princeton, NJ: Princeton University Press.

Avery, Robert K., and Thomas A. McCain. (1986). "Interpersonal and Mediated Encounters: A Reorientation to the Mass Communication Process." In Gary Gumpert and Robert Cathcart, Eds., *Inter/Media: Interpersonal Communication in a Media World* (pp. 121–131). New York: Oxford University Press.

Ayensu, Edward S. (1982). *The Timetable of Technology: A Record of the 20th Century's Amazing Achievements.* New York: Hearst Books.

Bagdikian, Ben H. (1971). *The Information Machines: Their Impact on Men and the Media.* New York: Harper & Row.

Bailey, Elaine K., and Morton Cotlar. (1994). "Teaching Via the Internet," *Communication Education, 43,* pp. 184–193.

Balazs, Bela. (1970). *Theory of the Film: Character and Growth of a New Art.* New York: Dover.

Baldwin, Charles Sears. (1959). *Ancient Rhetoric and Poetic Interpreted from Representative Works.* Gloucester, MA: Smith.

Barnes, Sue, and Leonore M. Greller. (1994). "Computer-Mediated Communication in the Organization," *Communication Education, 43,* pp. 129–142.

Batra, N. D. (1990). *A Self-Renewing Society: The Role of Television and Communications Technology.* Lanham, MD: University Press of America.

Becker, Lee B. (1980). "Measures of Gratification." In G. Cleveland Wilhoit and Harold deBock, Eds., *Mass Communication: Review Yearbook* (Vol. 1, pp. 229–248). Beverly Hills, CA: Sage.

Becker, S.W., A. Bavelas, & M. Braden. (1961). "An Index to Measure Contingency of English Sentences," *Language and Speech, 4,* pp. 138–145.

Bennett, W. Lance. (1992). "White Noise: The Perils of Mass Mediated Democracy," *Communication Monographs, 59,* pp. 401–406.

Benson, Thomas W. (1980). "The Rhetorical Structure of Frederick Wiseman's *High School,*" *Communication Monographs, 47,* pp. 233–261.

Benson, Thomas W., Ed. (1985). *Speech Communication in the 20th Century.* Carbondale: Southern Illinois University Press.

Benson, Thomas W. (1989). *American Rhetoric: Context and Criticism.* Carbondale: Southern Illinois University Press.

Benson, Thomas W. (1992). "Communication and the Circle of Learning," *Quarterly Journal of Speech, 78,* pp. 238–275.

Benson, Thomas W. (1994). "Electronic-Network Resources for Communication Scholars," *Communication Education, 43,* pp. 120–128.

Benson, Thomas W., and Carolyn Anderson. (1989). *Reality Fictions: The Films of Frederick Wiseman.* Carbondale: Southern Illinois University Press.

Benson, Thomas W., and Kenneth D. Frandsen. (1982). *Nonverbal Communication* (2nd ed.). Chicago: Science Research Associates.

Berg, David M. (1972). "Rhetoric, Reality, and Mass Media," *Quarterly Journal of Speech, 58,* pp. 255–263.

Berge, Zane L. (1994). "Electronic Discussion Groups," *Communication Education, 43,* pp. 102–111.

Berger, David G., and Richard A. Peterson. (1975). "Cycles in Symbolic Production: The Case of Popular Music," *American Sociological Review, 40,* pp. 158–173.

Berger, Peter L., and Thomas Luckmann. (1967). *The Social Construction of Reality: A Treatise in the Sociology of Knowledge*. New York: Doubleday.

Berlo, David K. (1960). *The Process of Communication: An Introduction to Theory and Practice*. New York: Holt, Rinehart & Winston.

Bertalanffy, Ludwig von. (1968). *General Systems Theory: Foundations, Development, Applications*. New York: Braziller.

Bertelsen, Dale A. (1992). "Media Form and Government: Democracy as an Archetypal Image in the Electronic Age," *Communication Quarterly*, 40, pp. 325–337.

Biltereyst, Daniel. (1991). "Resisting American Hegemony: A Comparative Analysis of the Reception of Domestic and US Fiction," *European Journal of Communication*, 6, pp. 469–497.

Biocca, Frank, Ed. (1991a). *Television and Political Advertising: Psychological Processes* (Vol. 1). Hillsdale, NJ: Erlbaum.

Biocca, Frank, Ed. (1991b). *Television and Political Advertising: Signs, Codes, and Images* (Vol. 2). Hillsdale, NJ: Erlbaum.

Biocca, Frank. (1992a, Autumn). "Communication Within Virtual Reality: Creating a Space for Research," *Journal of Communication*, 42, pp. 5–22.

Biocca, Frank. (1992b, Autumn). "Virtual Reality Technology: A Tutorial," *Journal of Communication*, 42, pp. 23–72.

Biskind, Peter. (1983). *Seeing is Believing: How Hollywood Taught Us to Stop Worrying and Love the Fifties*. New York: Pantheon Books.

Black, Edwin. (1965). *Rhetorical Criticism: A Study in Method*. New York: Macmillan.

Blair, Carole, and Mary L. Kahl. (1990). "Introduction: Revising the History of Rhetorical Theory," *Western Journal of Speech Communication*, 54, pp. 148–159.

Blankenship, Jane. (1966). *Public Speaking: A Rhetorical Perspective*. Englewood Cliffs, NJ: Prentice-Hall.

Blankenship, Jane. (1968). *A Sense of Style: An Introduction to Style for the Public Speaker*. Belmont, CA: Dickenson.

Bloodworth, John D. (1975). "Communication in the Youth Counter-Culture: Music as Expression," *Central States Speech Journal*, 26, pp. 304–309.

Bloome, David, and Ann Egan-Robertson. (1993). "The Social Construction of Intertextuality in Classroom Reading and Writing Lessons," *Reading Research Quarterly*, 28, pp. 304–333.

Blumler, Jay G. (1980). "The Role of Theory in Uses and Gratification Studies." In G. Cleveland Wilhoit and Harold deBock, Eds., *Mass Communication: Review Yearbook* (Vol. 1, pp. 201–228). Beverly Hills, CA: Sage.

Boase, Paul, Ed. (1980). *The Rhetoric of Protest and Reform, 1878–1898*. Athens: Ohio University Press.

Bogle, Donald. (1989). *Toms, Coons, Mulattoes, Mammies, and Bucks: An Interpretive History of Blacks in American Films* (Expanded ed.). New York: Continuum.

Bolter, Jay David. (1984). *Turing's Man: Western Culture in the Computer Age*. Chapel Hill: University of North Carolina Press.

Bolter, Jay David. (1991). *Writing Space: The Computer, Hypertext, and the History of Writing*. Hillsdale, NJ: Erlbaum.

Boorstin, Daniel J. (1961). *The Image: A Guide to Pseudo-Events in America*. New York: Harper & Row.

Booth, Mark W. (1976). "The Art of Words in Songs," *Quarterly Journal of Speech, 62*, pp. 242–249.

Bordwell, David. (1985). *Narration in the Fiction Film*. Madison: University of Wisconsin Press.

Bordwell, David, Janet Staiger, and Kristin Thompson. (1985). *The Classical Hollywood Cinema: Film Style and Mode of Production to 1960*. New York: Columbia University Press.

Bormann, Ernest G. (1965). *Theory and Research in the Communicative Arts*. New York: Holt, Rinehart & Winston.

Bormann, Ernest G. (1972). "Fantasy and Rhetorical Vision: The Rhetorical Criticism of Social Reality," *Quarterly Journal of Speech, 58*, pp. 396–407.

Boulding, Kenneth E. (1961). *The Image: Knowledge in Life and Society*. Ann Arbor: University of Michigan Press. (Original work published 1956)

Bove, Paul A. (1986). "The Ineluctability of Difference: Scientific Pluralism and the Critical Intelligence." In Jonathan Arac, Ed., *Postmodernism and Politics* (pp. 3–25). Minneapolis: University of Minnesota Press.

Braman, Sandra. (1993, Summer). "Harmonization of Systems: The Third Stage of the Information Society," *Journal of Communication, 43*, pp. 133–140.

Bramson, Leon, and Michael S. Schudson. (1987). "Mass Society." In *The New Encyclopaedia Britannica* (Vol. 16, pp. 940–944). Chicago: Encyclopaedia Britannica.

Bremmer, Jay, and Herman Roodenburg. (1991). *A Cultural History of Gesture*. Ithaca, NY: Cornell University Press.

Brigance, William Norwood, Ed. (1960). *A History and Criticism of American Public Address* (Vols. 1–2). New York: Russell & Russell. (Original work published 1943)

Brinton, Alan. (1990). "The Outmoded Psychology of Aristotle's *Rhetoric*," *Western Journal of Speech Communication, 54*, pp. 204–218.

Brock, Bernard L., and Robert L. Scott. (1980). "An Introduction to Rhetorical Criticism." In Bernard L. Brock and Robert L. Scott, Eds., *Methods of Rhetorical Criticism: A Twentieth-Century Perspective* (2nd ed., pp. 13–27). Detroit, MI: Wayne State University Press.

Brock, Bernard L., Robert L. Scott, and James W. Chesebro. (1990). "An Introduction to Rhetorical Criticism." In Bernard L. Brock, Robert L. Scott, and James W. Chesebro, Eds., *Methods of Rhetorical Criticism: A Twentieth-Century Perspective* (3rd ed., pp. 10–31). Detroit, MI: Wayne State University Press.

Brockriede, Wayne. (1974). "Rhetorical Criticism as Argument," *Quarterly Journal of Speech, 60*, pp. 165–174.

Brockriede, Wayne. (1985). "Constructs, Experience, and Argument," *Quarterly Journal of Speech, 71*, pp. 151–163.

Brown, Mary Ellen, Ed. (1990). *Television and Women's Culture: The Politics of the Popular*. Newbury Park, CA: Sage.

Brummett, Barry. (1992). *Toward a Discursive Ontology of Media*. Unpublished manuscript.

Bryant, Jennings, and Dolf Zillmann, Eds. (1991). *Responding to the Screen: Reception and Reaction Processes*. Hillsdale, NJ: Erlbaum.

Buettner-Janusch, John, and Michael H. Day. (1987). "Human Evolution." In *The New Encyclopaedia Britannica* (Vol. 18, pp. 930–980). Chicago: Encyclopaedia Britannica.

Burgoon, Judee, David Buller, and W. Gill Woodall. (1989). *Nonverbal Communication: The Unspoken Dialogue*. New York: Harper & Row.

Burgoon, Michael, Marshall Cohen, Michael D. Miller, and Charles L. Montgomery. (1978). "An Empirical Test of a Model of Resistance to Persuasion," *Human Communication Research, 5*, pp. 27–39.

Burke, Kenneth. (1965). *Permanence and Change: An Anatomy of Purpose*. Indianapolis, IN: Bobbs-Merrill. (Original work published 1935)

Burke, Kenneth. (1952a). "A Dramatistic View of the Origins of Language: Part One," *Quarterly Journal of Speech, 38*, 3, pp. 251–264.

Burke, Kenneth. (1952b). "A Dramatistic View of the Origins of Language: Part Two," *Quarterly Journal of Speech, 38*, 4, pp. 446–460.

Burke, Kenneth. (1953a). "A Dramatistic View of the Origins of Language: Part III," *Quarterly Journal of Speech, 39*, 1, pp. 79–92.

Burke, Kenneth. (1953b). "Postscripts on the Negative," *Quarterly Journal of Speech, 39*, 2, pp. 209–216.

Burke, Kenneth. (1970). *The Rhetoric of Religion: Studies in Logology*. Berkeley: University of California Press. (Original work published 1961)

Burleson, Brant R., and Michael S. Waltman. (1988). "Cognitive Complexity: Using the Role Category Questionnaire Measure." In Charles H. Tardy, Ed., *A Handbook for the Study of Human Communication: Methods and Instruments for Observing, Measuring, and Assessing Communication Processes* (pp. 1–35). Norwood, NJ: Ablex.

Burns, Edward McNall. (1963). *Western Civilizations: Their History and Their Culture* (6th ed.). New York: Norton.

Campbell, Gregg M. (1975). "Bob Dylan and the Pastoral Apocalypse," *Journal of Popular Culture, 8*, 696–707.

Campbell, Karlyn Kohrs. (1974). "Criticism: Ephemeral and Enduring," *Speech Teacher, 23*, pp. 9–14.

Campbell, Karlyn Kohrs. (1979). "The Nature of Criticism in Rhetorical and Communicative Studies," *Central States Speech Journal, 30*, pp. 4–13.

Campbell, Karlyn Kohrs. (1986). "Style and Content in the Rhetoric of Early Afro-American Feminists," *Quarterly Journal of Speech, 72*, pp. 434–445.

Campbell, Karlyn Kohrs. (1989a). *Man Cannot Speak for Her: A Critical Study of Early Feminist Rhetoric* (Vol. 1). New York: Greenwood Press.

Campbell, Karlyn Kohrs. (1989b). *Man Cannot Speak for Her: A Critical Study of Early Feminist Rhetoric* (Vol. 2). New York: Greenwood Press.

Campbell, Karlyn Kohrs, and Kathleen Hall Jamieson. (1978). "Form and Genre in Rhetorical Criticism: An Introduction." In Karlyn Kohrs Campbell and Kathleen Hall Jamieson, Eds., *Form and Genre: Shaping Rhetorical Action* (pp. 9–32). Falls Church, VA: Speech Communication Association.

Campbell, Warren, and Jack Heller. (1981). "Psychomusicology and Psycholinguistics: Parallel Paths or Separate Ways," *Psychomusicology, 1,* pp. 3–14.

Cantor, Muriel G., and Suzanne Pingree. (1983). *The Soap Opera.* Beverly Hills, CA: Sage.

Carey, James T. (1969a). "Changing Courtship Patterns in the Popular Song," *American Journal of Sociology, 74,* pp. 720–731.

Carey, James T. (1969b). "The Ideology of Autonomy in Popular Lyrics: A Content Analysis," *Psychiatry, 32,* pp. 150–164.

Carey, James W. (1989). *Communication as Culture: Essays on Media and Society.* Boston: Unwin Hyman.

Carpenter, Edmund. (1960). "The New Languages." In Edmund Carpenter and Marshall McLuhan, Eds., *Explorations in Communication: An Anthology* (pp. 162–179). Boston: Beacon Press.

Carpenter, Edmund, and Marshall McLuhan, Eds. (1960). *Explorations in Communication: An Anthology.* Boston: Beacon Press.

Carter, Bill. (1991, July 8). "TV Industry Unfazed by Rise in 'Zapping.' " *New York Times,* pp. D1, D6.

Carveth, Rodney, and Alison Alexander. (1985). "Soap Opera Viewing Motivations and the Cultivation Process," *Journal of Broadcasting and Electronic Media, 29,* pp. 259–273.

Cassirer, Ernst. (1965). *An Essay on Man: An Introduction to a Philosophy of Human Culture.* New Haven, CT: Yale University Press. (Original work published 1944)

Cathcart, Robert S. (1986). *Media Communities.* Unpublished manuscript.

Cathcart, Robert S. (1987, November). *Media Effects: Adding the Media Dimension to Rhetorical Criticism.* Paper presented at the annual meeting of the Speech Communication Association, Boston.

Cathcart, Robert, and Gary Gumpert (1983). "Mediated Interpersonal Communication: Towards a New Typology," *Quarterly Journal of Speech, 69,* pp. 267–277.

Chaffee, Steven H., and John L. Hochheimer. (1985). "The Beginnings of Political Communication Research in the United States: Origins of the 'Limited Effects Model.' " In Michael Gurevitch and Mark R. Levy, Eds., *Mass Communication Review Yearbook* (Vol. 5, pages 75–104). Beverly Hills, CA: Sage.

Chang, Briankle G. (1992). "Empty Intention," *Text and Performance Quarterly, 12,* pp. 212–227.

Chartier, Roger. (1987). *The Cultural Uses of Print in Early Modern France* (Lydia G. Cochrane, Trans.). Princeton, NJ: Princeton University Press.

Chenoweth, Lawrence. (1971). "The Rhetoric of Hope and Despair: A Study of the Jimi Hendrix Experience and the Jefferson Airplane," *American Quarterly, 23,* pp. 25–45.

Chesebro, James W. (1979). "Communication, Values, and Popular Television

Series—A Four Year Assessment." In Gary Gumpert and Robert Cathcart, Eds., *Inter/Media: Interpersonal Communication in a Media World* (pp. 528–560). New York: Oxford University Press.

Chesebro, James W. (1982). "Communication, Values and Popular Television Series—A Seven-Year Assessment." In Gary Gumpert and Robert Cathcart, Eds., *Inter/Media: Interpersonal Communication in a Media World* (2nd ed., pp. 468–519). New York: Oxford University Press.

Chesebro, James W. (1984). "The Media Reality: Epistemological Functions of Media in Cultural Systems," *Critical Studies in Mass Communication, 1*, pp. 111–130.

Chesebro, James W. (1985a). Definition as Rhetorical Strategy. *The Speech Communication Association of Pennsylvania Annual, 41*, pp. 5–15.

Chesebro, James W. (1985b, November). *Media Transformations: Revolutionary Challenges to the World's Cultures.* Paper presented at the annual meeting of the Speech Communication Association of Puerto Rico, San Juan.

Chesebro, James W. (1986a). "Communication, Values, and Popular Television Series—An Eleven Year Assessment." In Gary Gumpert and Robert Cathcart, Eds., *Inter/Media: Interpersonal Communication in a Media World* (3rd ed., pp. 477–512). New York: Oxford University Press.

Chesebro, James W. (1986b, December). *Media Transformations: Revolutionary Challenges to the World's Cultures, Part II.* Paper presented at the annual meeting of the Speech Communication Association of Puerto Rico, San Juan.

Chesebro, James W. (1986c). "Musical Patterns and Particular Musical Experiences," *Critical Studies in Mass Communication, 3*, pp. 256–260.

Chesebro, James W. (1987, December). *The Effects of Mass Media on Human Communication.* Public lecture presented at the University of Puerto Rico, Río Piedras, San Juan.

Chesebro, James W. (1989). "Text, Narration, and Media." *Text and Performance Quarterly, 9*, pp. 1–23.

Chesebro, James W. (1991). "Communication, Values, and Popular Television Series—A Seventeen-Year Assessment," *Communication Quarterly, 39*, pp. 197–225.

Chesebro, James W. (1995a). "Communication Technologies as Cognitive Systems." In Richard B. Gregg and Julia T. Wood, Eds., *Toward the Twenty-First Century: The Future of Speech Communication* (pp. 15–46). Cresskill, NJ: Hampton Press.

Chesebro, James W. (1995b, December 8). *Distinguishing Cultural Systems: Change as a Variable Explaining and Predicting Cross-Cultural Communication.* Keynote address at the annual meeting of the Speech Communication Association of Puerto Rico, San Juan.

Chesebro, James W., and Donald G. Bonsall. (1989). *Computer-Mediated Communication: Human Relationships in a Computerized World.* Tuscaloosa: University of Alabama Press.

Chesebro, James W., Davis A. Foulger, Jay E. Nachman, and Andrew Yannelli. (1985). "Popular Music as a Mode of Communication, 1955–1982," *Critical Studies in Mass Communication, 2*, pp. 115–135.

Chesebro, James W., and Caroline D. Hamsher. (1973). "Rhetorical Criticism: A Message-Centered Perspective," *Speech Teacher, 22*, pp. 282–290.

Chesebro, James W., and Kenneth L. Klenk. (1981). "Gay Masculinity in the Gay Disco." In James W. Chesebro, Ed., *Gayspeak: Gay Male and Lesbian Communication* (pp. 86–103, 325–330). New York: Pilgrim Press.

Christenson, Peter G., Peter DeBenedittus, and Thomas R. Lindlof. (1985). "Children's Use of Audio Media," *Communication Research, 12*, pp. 327–343.

Christenson, Peter G., and Jon Brian Peterson. (1988). "Genre and Gender in the Structure of Music Preferences," *Communication Research, 15*, 282–301.

Christianen, Michael. (1995). "Cycles in Symbol Production? A New Model to Explain Concentration, Diversity and Innovation in the Music Industry," *Popular Music, 14*, pp. 55–93.

Christians, Clifford G., Kim B. Rotzoll, and Mark Fackler. (1991). *Media Ethics: Cases and Moral Reasoning* (3rd ed.). New York: Longman.

Clanchy, M.T. (1979). *From Memory to Written Record: England, 1066–1307*. Cambridge, MA: Harvard University Press.

Clark, Donald Lemen. (1957). *Rhetoric in Greco-Roman Education*. New York: Columbia University Press.

Clarke, M.L. (1953). *Rhetoric at Rome: A Historical Survey*. New York: Barnes & Noble.

Clynes, Manfred. (1986). "Music Beyond the Score," *Communication and Cognition, 19*, pp. 169–194.

Cohen, Herman. (1985). "The Development of Research in Speech Communication: A Historical Perspective." In Thomas W. Benson, Ed., *Speech Communication in the 20th Century* (pp. 282–298). Carbondale: Southern Illinois University Press.

Cohen, Herman. (1994). *The History of Speech Communication: The Emergence of a Discipline, 1914–1945*. Annandale, VA: Speech Communication Association.

Cohen, Jodi R. (1991). "The 'Relevance' of Cultural Identity in Audiences' Interpretations of Mass Media," *Critical Studies in Mass Communication, 8*, pp. 442–454.

Collins, Catherine Ann, and Jeanne E. Clark. (1992). "A Structural Narrative Analysis of *Nightline's* 'This Week in the Holy Land,' " *Critical Studies in Mass Communication, 9*, pp. 25–43.

Collins, Mauri. (1994). "Internet Information Management Tools," *Communication Education, 43*, pp. 112–119.

Condit, Celeste Michelle. (1989). "The Rhetorical Limits of Polysemy," *Critical Studies in Mass Communication, 6*, pp. 103–122.

Corrigan, Timothy. (1991). *A Cinema Without Walls: Movies and Culture after Vietnam*. New Brunswick, NJ: Rutgers University Press.

Coulmas, Florian. (1989). *The Writing Systems of the World*. Cambridge, MA: Blackwell.

Cowser, R.L., Jr. (1978). "Uses of Antithesis in the Lyrics of Oscar Hammerstein II," *Journal of Popular Culture, 12*, pp. 365–374.

Cuddon, J.A. (1976). *A Dictionary of Literary Terms* (Rev. ed.). Garden City, NY: Doubleday.

Czitrom, Daniel J. (1982). *Media and the American Mind: From Morse to McLuhan*. Chapel Hill: University of North Carolina Press.

Dance, Frank E. X. (1969). "A Response to Gerald R. Miller's 'Human Information Processing: Some Research Guidelines.' " In Robert J. Kibler and Larry L. Barker, Eds., *Conceptual Frontiers in Speech-Communication: Report of the New Orleans Conference on Research and Instructional Development* (pp. 69–71). New York: Speech Association of America.

Dance, Frank E. X. (1989). "Ong's Voice: 'I,' the Oral Intellect, You, and We," *Text and Performance Quarterly*, 9, pp. 185–198.

Dance, Frank E. X., and Carl E. Larson. (1976). *The Functions of Human Communication: A Theoretical Approach*. New York: Holt, Rinehart & Winston.

Darnell, Donald K. (1969). "A Response to Gerald R. Miller's 'Human Information Processing: Some Research Guidelines.' " In Robert J. Kibler and Larry L. Barker, Eds., *Conceptual Frontiers in Speech-Communication: Report of the New Orleans Conference on Research and Instructional Development* (pp. 72–75). New York: Speech Association of America.

Darsey, James. (1981). "From 'Commies' and 'Queers' to 'Gay Is Good.' " In James W. Chesebro, Ed., *Gayspeak: Gay Male and Lesbian Communication* (pp. 224–247). New York: Pilgrim Press.

Darsey, James. (1991). "From 'Gay Is Good' to the Scourge of AIDS: The Evolution of Gay Liberation Rhetoric, 1977–1990," *Communication Studies*, 41, pp. 43–66.

DeFleur, Melvin L., and Sandra Ball-Rokeach. (1982). *Theories of Mass Communication* (4th ed.). New York: Longman.

DeFleur, Melvin L., and Sandra Ball-Rokeach. (1989). *Theories of Mass Communication* (5th ed.). New York: Longman.

Delia, Jesse G. (1987). "Communication Research: A History." In Charles R. Berger and Steven H. Chaffee, Eds., *Handbook of Communication Science* (pp. 20–98). Newbury Park, CA: Sage.

Delia, Jesse G., and Barbara J. O'Keefe. (1979). "Constructivism: The Development of Communication in Children." In Ellen Wartella, Ed., *Children Communicating: Media and the Development of Thought, Speech, Understanding* (pp. 157–185). Beverly Hills, CA: Sage.

Denny, J. Peter. (1991). "Rational Thought in Oral Culture and Literate Contextualization." In David R. Olson and Nancy Torrance, Eds., *Literacy and Orality* (pp. 66–89). New York: Cambridge University Press.

Derrida, Jacques. (1976). *Of Grammatology* (Gayatri Chakravorty Spivak, Trans.). Baltimore, MD: Johns Hopkins University Press. (Original work published 1967)

Desmond, Roger Jon. (1987). "Adolescents and Music Lyrics: Implications of a Cognitive Perspective," *Communication Quarterly*, 35, pp. 276–284.

Dizard, Wilson P., Jr. (1989). *The Coming Information Age: An Overview of Technology, Economics, and Politics* (3rd ed.). New York: Longman.

Doane, Mary Ann. (1985). "The Voice in the Cinema: The Articulation of Body and Space." In Bill Nichols, Ed., *Movies and Methods* (Vol. 2, pp. 565–576). Berkeley: University of California Press.

Donohue, Thomas R., and Timothy P. Meyer. (1984). "Children's Understanding of Television Commercials: The Acquisition of Competence." In Robert N. Bostrom, Ed., *Competence in Communication: A Multidisciplinary Approach* (pp. 129–149). Beverly Hills, CA: Sage.

Drucker, Susan J., and Gary Gumpert. (1991). "Public Space and Communication: The Zoning of Public Interaction," *Communication Theory, 1*, pp. 294–310.

Dunham, Barrows. (1975). "Of Philosophy, Love, and E. Y. Harburg," *Journal of Communication, 25*, pp. 69–73.

Durey, Jill Felicity. (1991). "The State of Play and Interplay in Intertextuality," *Style, 25*, pp. 616–635.

Eastman, Susan Tyler, and Gregory D. Newton. (1995, Winter). "Delineating Grazing: Observations of Remote Control Use," *Journal of Communication, 45*, pp. 77–95.

Eckhardt, Beverly B., Mary R. Wood, and Robin Smith Jacobvitz. (1991). "Verbal Ability and Prior Knowledge: Contributions to Adults' Comprehension of Television," *Communication Research, 18*, pp. 636–649.

Eco, Umberto. (1983). *The Name of the Rose* (William Weaver, Trans.). New York: Warner Books. (Original work published 1980)

Edelman, Murray. (1988). *Constructing the Political Spectacle*. Chicago: University of Chicago Press.

Edmonson, W. H. (1971). *Lore: An Introduction to the Science of Folklore and Literature*. New York: Holt, Rinehart & Winston.

Edwards, Vic, and Thomas S. Sienkewicz. (1990). *Oral Cultures Past and Present: Rappin' and Homer*. Cambridge, MA: Blackwell.

Ehninger, Douglas. (1968). "On Systems of Rhetoric," *Philosophy and Rhetoric, 1*, pp. 131–144.

Eisenson, Jon. (1974). *Voice and Diction: A Program for Improvement* (3rd ed.). New York: Macmillan.

Eisenstein, Elizabeth L. (1979). *The Printing Press as an Agent of Change: Communications and Cultural Transformations in Early-Modern Europe* (Vols. 1–2). New York: Cambridge University Press.

Eisenstein, Elizabeth L. (1980). "The Emergence of Print Culture in the West," *Journal of Communication, 30*, pp. 99–106.

Ekman, Paul, and W.V. Friesen. (1969). "The Repertoire of Nonverbal Behavior: Categories, Origins, Usage, and Coding," *Semiotica, 1*, pp. 49–98.

Elliott, Norman. (1979). "Language and Cognition in the Developing Child." In Ellen Wartella, Ed., *Children Communicating: Media and Development of Thought, Speech, Understanding* (pp. 187–214). Beverly Hills, CA: Sage.

Elliott-Faust, Darlene J., and Michael Pressley. (1986). "How to Teach Comparison Processing to Increase Children's Short- and Long-Term Listening Comprehension Monitoring," *Journal of Educational Psychology, 78*, pp. 27–33.

Ellis, Bill. (1978). "The 'Blind' Girl and the Rhetoric of Sentimental Heroism," *Journal of American Folklore, 91*, pp. 657–674.

"English Assignments in 'Hypertext.' " (1989, January 25). *The Chronicle of Higher Education*, p. A18.

Enos, Richard Leo, Ed. (1990). *Oral and Written Communication: Historical Approaches*. Newbury Park, CA: Sage.

Estep, Rhoda, and Patrick T. Macdonald. (1985). "Crime in the Afternoon: Murder and Robbery on Soap Operas," *Journal of Broadcasting and Electronic Media, 29*, pp. 323–331.

Faber, Ronald J., Jane D. Brown, and Jack McLeod. (1986). "Coming of Age in the Global Village: Television and Adolescence." In G. Gumpert and R. Cathcart (Eds.), *Inter/Media: Interpersonal Communication in a Media World* (3rd ed., pp. 550–572). New York: Oxford University Press.

Febvre, Lucien, and Henri-Jean Martin. (1990). *The Coming of the Book: The Impact of Printing 1450–1800* (David Gerard, Trans.). New York: Verso. (Original work published 1958)

Ferrell, Keith. (1988, February). "Hypertext: Here, There, and Everywhere," *Compute, 10*, pp. 26–28.

Fisher, Walter R. (1984). "Narration as a Human Communication Paradigm: The Case of Public Moral Argument," *Communication Monographs, 51*, pp. 1–22.

Fiske, John. (1986). "Television: Polysemy and Popularity," *Critical Studies in Mass Communication, 3*, pp. 391–408.

Foss, Sonja K. (1979). "The Equal Rights Amendment Controversy: Two Worlds in Conflict," *Quarterly Journal of Speech, 65*, pp. 275–288.

Foss, Sonja K. (1989). *Rhetorical Criticism: Exploration and Practice*. Prospect Heights, IL: Waveland Press.

Foss, Sonja K., Karen A. Foss, and Robert Trapp. (1985). *Contemporary Perspectives on Rhetoric*. Prospect Heights, IL: Waveland Press.

Foster, Hal. (1983). "Postmodernism: A Preface." In Hal Foster, Ed., *The Anti-Aesthetic: Essays on Postmodern Culture* (pp. ix–xvi). Port Townsend, WA: Bay Press.

Foster, Harold M. (1979). *The New Literacy: The Language of Film and Television*. Urbana, IL: National Council of Teachers of English.

Fowler, Elizabeth M. (1983, September 7). "Global Public Relations." *New York Times*, p. D20.

Fowles, Jib. (1992). *Why Viewers Watch: A Reappraisal of Television's Effects*. Newbury Park, CA: Sage.

Freedman, David. (1994). *Brainmakers: How Scientists Are Moving Beyond Computers to Create a Rival to the Human Being*. New York: Simon & Schuster.

Furno-Lamude, Diane, and James Anderson. (1992). "The Uses and Gratifications of Rerun Viewing," *Journalism Quarterly, 69*, pp. 362–372.

Gardner, Howard. (1983). *Frames of Mind: The Theory of Multiple Intelligences*. New York: Basic Books.

Garnham, Nicholas. (1990). *Capitalism and Communication: Global Culture and the Economics of Information*. Newbury Park, CA: Sage.

Gelb, Ignace J. (1987). "Writing." In *The New Encyclopaedia Britannica* (Vol. 29, pp. 982–990). Chicago: Encyclopaedia Britannica.

Gelernter, David. (1994a). *The Muse in the Machine: Computerizing the Poetry of Human Thought.* New York: Free Press.

Gelernter, David. (1994b, September 19, 26). "Unplugged: The Myth of Computers in the Classroom," *The New Republic*, pp. 14–15.

Gerbner, George, and Larry Gross. (1976, Spring). "Living with Television: The Violence Profile," *Journal of Communication, 26*, pp. 172–199.

Gerbner, George, Larry Gross, Michael F. Eleey, Marilyn Jackson-Beeck, Suzanne Jeffries-Fox, and Nancy Signorielli. (1977, Spring). "TV Violence Profile No. 8: The Highlights." *Journal of Communication, 27*, pp. 171–180.

Gerbner, George, Larry Gross, Michael Morgan, and Nancy Signorielli. (1982, Spring). "Charting the Mainstream: Television's Contribution to Political Orientation." *Journal of Communication, 32*, pp. 100–127.

Gergen, Kenneth J. (1991). *The Saturated Self: Dilemmas of Identity in Contemporary Life.* New York: Basic Books.

Gfeller, Kate E. (1990a). "Cultural Context as it Relates to Music Therapy." In Robert F. Unkefer, Ed., *Music Therapy in the Treatment of Adults with Mental Disorders: Theoretical Bases and Clinical Interventions* (pp. 63–69). New York: Schirmer Books.

Gfeller, Kate E. (1990b). "The Function of Aesthetic Stimuli in the Therapeutic Process." In Robert F. Unkefer, Ed., *Music Therapy in the Treatment of Adults with Mental Disorders: Theoretical Bases and Clinical Interventions* (pp. 70–81). New York: Schirmer Books.

Gfeller, Kate E. (1990c). "Music as Communication." In Robert F. Unkefer, Ed., *Music Therapy in the Treatment of Adults with Mental Disorders: Theoretical Bases and Clinical Interventions* (pp. 50–62). New York: Schirmer Books.

Giannetti, Louis D. (1976). *Understanding Movies* (2nd ed.). Englewood Cliffs, NJ: Prentice-Hall.

Gilder, George. (1989). *Microcosm: The Quantum Revolution in Economics and Technology.* New York: Simon & Schuster.

Goffman, Erving. (1959). *The Presentation of Self in Everyday Life.* Garden City, NY: Doubleday.

Goldblatt, Harold C. (1987). "Chinese Language." In *The New Encyclopaedia Britannica* (Vol. 22, pp. 256–265). Chicago: Encyclopaedia Britannica.

Gonzalez, Alberto, and John J. Makay. (1987). "Rhetorical Ascription and the Gospel According to Dylan," *Quarterly Journal of Speech, 69*, pp. 1–14.

Goodenough, Ward H. (1971). *Culture, Language, and Society.* Reading, MA: Addison-Wesley.

Goody, Jack. (1986). *The Logic of Writing and the Organization of Society.* New York: Cambridge University Press.

Goody, Jack, and Ian Watt. (1963). "The Consequences of Literacy," *Comparative Studies in Society and History, 5*, pp. 304–345.

Gordon, George N. (1971). *Persuasion: The Theory and Practice of Manipulative Communication.* New York: Hastings House.

Gorman, Michael E., and W. Bernard Carlson. (1990, Spring). "Interpreting Invention as a Cognitive Process: The Case of Alexander Graham Bell,

Thomas Edison, and the Telephone," *Science, Technology, and Human Values*, 15, pp. 131–164.

Gow, Joe. (1994). "Mood and Meaning in Music Video: The Dynamics of Audio-Visual Synergy," *Southern Communication Journal*, 59, pp. 255–261.

Gozzi, Raymond, Jr., and W. Lance Haynes. (1992). "Electric Media and Electric Epistemology: Empathy at a Distance," *Critical Studies in Mass Communication*, 9, pp. 217–228.

Grafton, Carl, and Anne Permaloff. (1991, December). "Hypertext and Hypermedia," *PS: Political Science and Politics*, 24, pp. 724–730.

Greenberg, Bradley S., and Dave D'Alessio. (1985). "Quantity and Quality of Sex in the Soaps," *Journal of Broadcasting and Electronic Media*, 29, pp. 309–321.

Greene, John O. (1988). "Cognitive Processes: Methods for Probing the Black Box." In Charles H. Tardy, Ed., *A Handbook for the Study of Human Communication: Methods and Instruments for Observing, Measuring, and Assessing Communication Processes* (pp. 37–66). Norwood, NJ: Ablex.

Greenfield, Patricia Marks. (1993). "Representational Competence in Shared Symbol Systems: Electronic Media from Radio to Video Games." In Rodney R. Cocking and K. Ann Renninger, Eds., *The Development and Meaning of Psychological Distance* (pp. 161–183). Hillsdale, NJ: Erlbaum.

Greenfield, Patricia, and Jessica Beagles-Roos. (1988, Spring). "Radio vs. Television: Their Cognitive Impact on Children of Different Socioeconomic and Ethnic Groups," *Journal of Communication*, 38, pp. 71–92.

Gregg, Richard, B. (1984). *Symbolic Inducement and Knowing: A Study in the Foundations of Rhetoric*. Columbia: University of South Carolina Press.

Grodin, Debra. (1991). "The Interpreting Audience: The Therapeutics of Self-Help Book Reading," *Critical Studies in Mass Communication*, 8, pp. 404–420.

Gross, Larry. (1974). "Modes of Communication and the Acquisition of Symbolic Competence." In David R. Olson, Ed., *Media and Symbols: The Forms of Expression, Communication, and Education* (pp. 56–80). Chicago: National Society for the Study of Education.

Grossberg, Lawrence. (1986). "Is There Rock after Punk?" *Critical Studies in Mass Communication*, 3, pp. 50–74.

Grossberg, Lawrence. (1988). "Putting the Pop Back into Postmodernism." In Andrew Ross, Ed., *Universal Abandon? The Politics of Postmodernism* (pp. 167–190). Minneapolis: University of Minnesota Press.

Grun, Bernard. (1982). *The Timetables of History: A Horizontal Linkage of People and Events*. New York: Simon & Schuster.

Gumpert, Gary. (1970, September). "The Rise of Mini-Comm," *Journal of Communication*, 20, pp. 280–290.

Gumpert, Gary. (1975, Fall). "The Rise of Uni-Comm," *Today's Speech*, 23, pp. 34–38.

Gumpert, Gary. (1987). *Talking Tombstones and Other Tales of the Media Age*. New York: Oxford University Press.

Gumpert, Gary, and Robert Cathcart. (1985). "Media Grammars, Generations, and Media Gaps," *Critical Studies in Mass Communication*, 2, pp. 23–35.

Hacker, Kenneth L., and Tara G. Coste. (1992). "A Political Linguistics Analysis of Network Television News Viewers' Discourse," *The Howard Journal of Communications*, 3, pp. 299–316.

Hall, Edward T. (1969). *The Hidden Dimension*. Garden City, NY: Doubleday.

Hall, James W. (1968). "Concepts of Liberty in American Broadside Ballads," *Journal of Popular Culture*, 2, pp. 252–275.

Harrington, John. (1973). *The Rhetoric of Film*. New York: Holt, Rinehart & Winston.

Harris, Richard Jackson. (1994). *A Cognitive Psychology of Mass Communication* (2nd ed.). Hillsdale, NJ: Erlbaum.

Harris, Roy. (1986). *The Origins of Writing*. La Salle, IL: Open Court.

Hartman, Douglas K. (1991). "The Intertextual Links of Readers Using Multiple Passages: A Postmodern/Semiotic/Cognitive View of Meaning Making," *National Reading Conference Yearbook*, 40, pp. 49–66.

Hastings, Arthur. (1970). "Metaphor in Rhetoric," *Western Journal of Speech Communication*, 34, pp. 181–194.

Havelock, Eric A. (1963). *Preface to Plato*. Cambridge, MA: Belknap Press of Harvard University Press.

Havelock, Eric A. (1980). "The Coming of Literate Communication to Western Culture," *Journal of Communication*, 30, pp. 90–98.

Havelock, Eric A. (1982). *The Literate Revolution in Greece and Its Cultural Consequences*. Princeton, NJ: Princeton University Press.

Havelock, Eric A. (1986). *The Muse Learns to Write: Reflections on Orality and Literacy from Antiquity to the Present*. New Haven, CT: Yale University Press.

Hawken, Paul. (1983). *The Next Economy*. New York: Holt, Rinehart & Winston.

Hawkes, Terence. (1977). *Structuralism and Semiotics*. Berkeley: University of California Press.

Haynes, W. Lance. (1988). "Of That Which We Cannot Write: Some Notes on the Phenomenology of Media," *Quarterly Journal of Speech*, 74, pp. 71–101.

Henley, Nancy M. (1977). *Body Politics: Power, Sex, and Nonverbal Communication*. Englewood Cliffs, NJ: Prentice-Hall.

Herbener, Gerald F., G. Norman Tubergen, and S. Scott Whitlow. (1979). "Dynamics of the Frame in Visual Composition," *Educational Technology and Communications Journal*, 27, pp. 83–88.

Hochmuth, Marie Kathryn, Ed. (1955). *A History and Criticism of American Public Address* (Vol. 3). New York: Longmans, Green.

Holmberg, Carl Bryan. (1985). "Toward a Rhetoric of Music: 'Dixie,'" *Southern Speech Communication Journal*, 51, pp. 71–82.

Homer. (1990). *The Iliad* (Robert Fagles, Trans.). New York: Penguin Books.

Horton, Donald, and R. Richard Wohl. (1956). "Mass Communication and Para-Social Interaction: Observation on Intimacy at a Distance," *Psychiatry*, 19, pp. 215–229.

Hughes, Thomas P. (1983). *American Genesis: A Century of Innovation and Technological Enthusiasm, 1870–1970*. New York: Penguin Books.

Huizinga, Johan. (1954). *The Waning of the Middle Ages: A Study of the Forms of Life, Thought and Art in France and the Netherlands in the XIVth and XVth Centuries*. Garden City, NY: Doubleday.

Illich, Ivan. (1991). "A Plea for Research on Lay Literacy." In David R. Olson and Nancy Torrance, Eds., *Literacy and Orality* (pp. 28–46). New York: Cambridge University Press.

Illich, Ivan, and Barry Sanders. (1988). *The Alphabetization of the Popular Mind*. San Francisco, CA: North Point Press.

Innis, Harold. (1951). *The Bias of Communication*. Toronto: Toronto University Press.

International Encyclopedia of Communications. (1989). New York: Oxford University Press.

Irvine, James R., and Walter G. Kirkpatrick. (1972). "The Musical Form in Rhetorical Exchange," *Quarterly Journal of Speech, 58*, pp. 272–289.

Jackendoff, Ray. (1992). "Musical Processing and Musical Affect." In Mari Riess Jones and Susan Holleran, Eds., *Cognitive Bases of Musical Communication* (pp. 51–68). Washington, DC: American Psychological Association.

Jacoby, Jacob, and Wayne D. Hoyer. (1982a). "On Miscomprehending Televised Communication: A Rejoinder," *Journal of Marketing, 46*, pp. 35–43.

Jacoby, Jacob, and Wayne D. Hoyer. (1982b). "Viewer Miscomprehension of Televised Communication," *Journal of Marketing, 46*, pp. 12–26.

Jacoby, Jacob, and Wayne D. Hoyer. (1987). *The Comprehension and Miscomprehension of Print Communications: An Investigation of Mass Media Magazines*. New York: Advertising Educational Foundation.

Jacoby, Jacob, Wayne D. Hoyer, and David A. Sheluga. (1980). *The Comprehension and Miscomprehension of Televised Communication*. New York: American Association of Advertising Agencies.

James, William. (1950). *The Principles of Psychology* (Vol. 2). New York: Dover/Henry Holt. (Original work published 1869)

Jamieson, Kathleen Hall. (1988). *Eloquence in an Electronic Age: The Transformation of Political Speechmaking*. New York: Oxford University Press.

Jandt, Fred J. (1995). *Intercultural Communication: An Introduction*. Thousand Oaks, CA: Sage.

Jasinski, James. (1992). "Rhetoric and Judgment in the Constitutional Ratification Debate of 1787–1788: An Exploration in the Relationship between Theory and Critical Practice," *Quarterly Journal of Speech, 78*, pp. 197–218.

Jaynes, Julian. (1976). *The Origins of Consciousness in the Breakdown of the Bicameral Mind*. Boston: Houghton Mifflin.

Johnson, Nan. (1991). *Nineteenth-Century Rhetoric in North America*. Carbondale: Southern Illinois University Press.

Jones, Steven. (1993). "A Sense of Space: Virtual Reality, Authenticity and the Aural," *Critical Studies in Mass Communication, 10*, pp. 238–252.

Joyce, Robert. (1975). *The Esthetic Animal: Man, the Art-Created Art Creator*. Hicksville, NY: Exposition Press.

Kaufer, David S., and Kathleen M. Carley. (1993). *Communication at a Distance: The Influence of Print on Sociocultural Organization and Change*. Hillsdale, NJ: Erlbaum.

Kawin, Bruce F. (1987). *How Movies Work*. New York: Macmillan.

Kellner, Douglas. (1995, May). "No Respect! Disciplinarity and Media Studies in

Communication. Media Communications vs. Cultural Studies: Overcoming the Divide," *Communication Theory, 5,* pp. 162–177.

Kennedy, George. (1963). *The Art of Persuasion in Greece.* Princeton, NJ: Princeton University Press.

Kibler, Robert J., and Larry L. Barker, Eds. (1969). *Conceptual Frontiers in Speech-Communication: Report of the New Orleans Conference on Research and Instructional Development.* New York: Speech Association of America.

Klapper, Joseph T. (1960). *The Effects of Mass Communication.* New York: Free Press.

Knapp, Mark L., and Judith A. Hall. (1992). *Nonverbal Communication in Human Interaction* (3rd ed.). New York: Holt, Rinehart & Winston.

Kneale, Dennis. (1988, April 25). " 'Zapping' of TV Ads Appears Pervasive," *The Wall Street Journal,* p. 29.

Knox, Bernard. (1990). "Introduction [to] *The Iliad.*" In Homer, *The Iliad* (Robert Fagles, Trans.; pp. 3–64). New York: Penguin Books.

Knupp, Ralph E. (1981). "A Time for Every Purpose Under Heaven: Rhetorical Dimensions of Protest Music," *Southern Speech Communication Journal, 46,* pp. 377–389.

Kochman, Thomas. (1972). "Towards an Ethnography of Black American Speech Behavior." In Thomas Kochman, Ed., *Rappin' and Stylin' Out: Communication in Urban Black America* (pp. 241–264). Urbana: University of Illinois Press.

Komatsuzaki, Seisuke. (1981). "The Impact of Communication Theory on the Evolution of Media," *International Social Science Journal, 33,* pp. 91–98.

Kosokoff, Stephen, and Carl W. Carmichael. (1970). "The Rhetoric of Protest: Song, Speech, and Attitude Change," *Southern Speech Communication Journal, 35,* pp. 295–302.

Kraft, Robert N., Phillip Cantor, and Charles Gottdiener. (1991). "The Coherence of Visual Narratives," *Communication Research, 18,* pp. 601–616.

Kraut, Robert. (1992). "On the Possibility of a Determinate Semantics for Music." In Mari Riess Jones and Susan Holleran, Eds., *Cognitive Bases of Musical Communication* (pp. 11–22). Washington, DC: American Psychological Association.

Krippendorff, Klaus. (1984). "An Epistemological Function for Communication," *Journal of Communication, 34,* pp. 21–36.

Kroeber, Alfred L., and Clyde Kluckhohn, with the assistance of Wayne Untereiner and appendices by Alfred G. Meyer. (1952). *Culture: A Critical Review of Concepts and Definitions* (Vol. 47, No. 1). Cambridge, MA: Papers of the Peabody Museum of American Archaeology and Ethnology/Harvard University Press.

Krugman, Herbert E. (1965). "The Impact of Television Advertising: Learning Without Involvement," *Public Opinion Quarterly, 29,* pp. 349–356.

Krugman, Herbert E. (1971). "Brain Wave Measures of Media Involvement," *Journal of Advertising Research, 11,* pp. 3–9.

Krumhansl, Carol L. (1992). "Internal Representations for Music Perception and Performance." In Mari Riess Jones and Susan Holleran, Eds., *Cognitive Bases of Musical Communication* (pp. 197–212). Washington, DC: American Psychological Association.

Kuehn, Scott A. (1994). "Computer-Mediated Communication in Instructional Settings: A Research Agenda," *Communication Education, 43*, pp. 171–183.

Lanham, Richard A. (1993). *The Electronic Word: Democracy, Technology, and the Arts*. Chicago: University of Chicago Press.

Lanier, Jaron, and Frank Biocca. (1992, Autumn). "An Insider's View of the Future of Virtual Reality," *Journal of Communication, 42*, pp. 150–172.

LeCoat, Gerard G. (1976). "Music and the Three Appeals of Classical Rhetoric," *Quarterly Journal of Speech, 62*, pp. 157–166.

Lemert, James B. (1989). *Criticizing the Media: Empirical Approaches*. Newbury Park, CA: Sage.

Lentz, Tony M. (1989). *Orality and Literacy in Hellenic Greece*. Carbondale: Southern Illinois University Press.

Lessac, Arthur. (1967). *The Use and Training of the Human Voice: A Practical Approach to Speech and Voice Dynamics* (2nd ed.). New York: Drama Book.

Levy, Mark R. (1985). "VCR Use and the Concept of Audience Activity," *Communication Quarterly, 35*, pp. 267–275.

Lewis, George H. (1976). "Country Music Lyrics," *Journal of Communication, 26*, 37–40.

Lewis, Peter H. (1994, June 14). "Virtual Reality Plans to Grow More Real," *New York Times*, p. B8.

Liebert, Robert M., Joyce N. Sprafkin, and Emily S. Davidson. (1982). *The Early Window: Effects of Television on Children and Youth* (2nd ed.). New York: Pergamon Press.

Liebes, Tamar. (1988). "Cultural Differences in the Retelling of Television Fiction," *Critical Studies in Mass Communication, 5*, pp. 277–292.

Lindesmith, Alfred R., and Anselm L. Strauss. (1956). *Social Psychology* (Rev. ed.). New York: Holt, Rinehart & Winston.

Lippmann, Walter. (1925). *The Phantom Public*. New York: Harcourt, Brace.

Logan, Robert K. (1986). *The Alphabet Effect: The Impact of the Phonetic Alphabet on the Development of Western Civilization*. New York: Morrow.

Lord, Albert B. (1960). *The Singer of Tales*. Cambridge, MA: Harvard University Press.

Lull, James. (1985). "On the Communicative Properties of Music," *Communication Research, 12*, pp. 363–372.

Luria, Alexander Romanovich. (1976). *Cognitive Development: Its Cultural and Social Foundations*. Cambridge, MA: Harvard University Press.

Lustig, Myron W., and Jolene Koester. (1996). *Intercultural Competence: Interpersonal Communication Across Cultures* (2nd ed.). New York: HarperCollins.

Lutz, Jeanne, Robert B. Huber, Carroll C. Arnold, John F. Wilson, and James W. Chesebro. (1985). "From the 50th to the 75th: ECA History through the Eyes of Past Presidents," *Communication Quarterly, 33*, pp. 3–16.

Lyotard, Jean-Francois. (1979). *The Postmodern Condition: A Report on Knowledge* (Geoff Bennington and Brian Massumi, Trans.). Minneapolis: University of Minnesota Press.

MacKenzie, Donald, and Judy Wajcman, Eds. (1985). *The Social Shaping of Technology*. Milton Keynes, UK/Philadelphia: Open University Press.

Maheau, Rene. (1972). Quoted in John McHale, Ed., *World Facts and Trends: Where Man Is Headed—A Multi-Dimensional View* (p. 83). New York: Macmillan.

Makus, Ann. (1990). "Rhetoric Then and Now: A Proposal for Integration," *Western Journal of Speech Communication, 54*, pp. 189–203.

Manchester, William. (1992). *A World Lit Only By Fire: The Medieval Mind and the Renaissance, Portrait of an Age*. Boston: Little, Brown.

Mander, Jerry. (1978). *Four Arguments for the Elimination of Television*. New York: Morrow.

Marc, David. (1984). *Demographic Vistas: Television in American Culture*. Philadelphia: University of Pennsylvania Press.

Marvin, Carolyn. (1985). "Space, Time, and Captive Communication History." In Michael Gurevitch and Mark R. Levy, Eds., *Mass Communication Review Yearbook* (Vol. 5, pp. 105–124). Beverly Hills, CA: Sage.

Marx, Leo. (1964). *The Machine in the Garden: Technology and the Pastoral Ideal in America*. New York: Oxford University Press.

Maurer, D., and Maurer, C. (1988). *The World of the Newborn*. New York: Basic Books.

McComb, Mary. (1994). "Benefits of Computer-Mediated Communication in College Courses," *Communication Education, 43*, pp. 159–170.

McCroskey, James C. (1993). *An Introduction to Rhetorical Communication* (6th ed.). Englewood Cliffs, NJ: Prentice-Hall.

McDonald, Daniel G., and Russell Schechter. (1988). "Audience Role in the Evolution of Fictional Television Content," *Journal of Broadcasting and Electronic Media, 32*, pp. 61–71.

McGuire, Michael. (1984). " 'Darkness on the Edge of Town': Bruce Springsteen's Rhetoric of Optimism and Despair." In Martin J. Medhurst and Thomas W. Benson, Eds., *Rhetorical Dimensions in Media: A Critical Casebook* (pp. 233–250). Dubuque, IA: Kendall/Hunt.

McHale, John. (1972). *World Facts and Trends: Where Man Is Headed—A Multi-Dimensional View*. New York: Macmillan.

McLaughlin, Lisa. (1995, May). "No Respect! Disciplinarity and Media Studies in Communication. Feminist Communication Scholarship and 'The Women Question' in the Academy," *Communication Theory, 5*, pp. 144–161.

McLuhan, Marshall. (1962). *The Gutenberg Galaxy: The Making of Typographic Man*. Toronto: University of Toronto Press.

McLuhan, Marshall. (1964). *Understanding Media: The Extensions of Man*. New York: New American Library.

McLuhan, Marshall, and Bruce R. Powers. (1989). *The Global Village: Transformations in World Life and Media in the 21st Century*. New York: Oxford University Press.

McQuail, Denis, Ed. (1972). *Sociology of Mass Communications*. New York: Penguin Books.

McQuail, Denis. (1980). "The Historicity of Mass Media Science." In G. Cleveland Wilhoit and Harold de Bock, Eds., *Mass Communication Review Yearbook* (Vol. 1, pages 109–123). Beverly Hills, CA: Sage.

McQuail, Denis. (1984). "With the Benefits of Hindsight: Reflections on Uses and Gratifications Research," *Critical Studies in Mass Communication, 1*, pp. 177–193.

McQuail, Denis. (1987). *Mass Communication Theory: An Introduction* (2nd ed.). Beverly Hills, CA: Sage.

Medhurst, Martin J. (1982). "*Hiroshima, Mon Amour:* From Iconography to Rhetoric," *Quarterly Journal of Speech, 68*, pp. 345–370.

Medhurst, Martin J. (1993). "The Rhetorical Structure of Oliver Stone's *JFK,*" *Critical Studies in Mass Communication, 10*, pp. 128–143.

Medhurst, Martin J., and Thomas W. Benson, Eds. (1984). *Rhetorical Dimensions in Media: A Critical Casebook*. Dubuque, IA: Kendall/Hunt.

"Media Critics" [Special issue]. (1995, Spring). Media *Studies Journal, 9*, p. 2.

Mehrabian, Albert. (1968, February). "Communication Without Words." *Psychology Today, 2*, pp. 53–55.

Messaris, Paul. (1994). "Does TV Belong in the Classroom? Cognitive Consequences of Visual 'Literacy.' " In Stanley A. Deetz, Ed., *Communication Yearbook* (Vol. 17, pp. 431–452). Thousand Oaks, CA: Sage.

Meyer, L.B. (1956). *Emotion and Meaning in Music*. Chicago: University of Chicago Press.

Meyrowitz, Joshua. (1985). *No Sense of Place: The Impact of Electronic Media on Social Behavior*. New York: Oxford University Press.

Miller, Carolyn R. (1978). "Technology as a Form of Consciousness: A Study of Contemporary Ethos," *Central States Speech Journal, 29*, pp. 228–236.

Miller, Gerald R. (1969). "Human Information Processing: Some Research Guidelines." In Robert J. Kibler and Larry L. Barker, Eds., *Conceptual Frontiers in Speech-Communication: Report of the New Orleans Conference on Research and Instructional Development* (pp. 51–68). New York: Speech Association of America.

Miller, Gerald. (1986). "A Neglected Connection: Mass Media Exposure and Interpersonal Communicative Competency." In Gary Gumpert and Robert Cathcart, Eds., *Inter/Media: Interpersonal Communication in a Media World* (3rd ed., pp. 132–139). New York: Oxford University Press.

Millerson, Gerald. (1982). *The Technique of Lighting for Television and Motion Pictures* (2nd ed.). London: Focal Press.

Misa, Thomas J. (1992, Winter). "Theories of Technological Change: Parameters and Purposes," *Science, Technology, and Human Values, 17*, pp. 3–12.

Mohrmann, G.P., and F. Eugene Scott. (1976). "Popular Music and World War II: The Rhetoric of Continuation," *Quarterly Journal of Speech, 62*, pp. 145–156.

Molloy, J. T. (1975). *Dress for Success*. New York: Warner Books.

Monaco, James. (1981). *How to Read a Film: The Art, Technology, Language, History, and Theory of Film and Media* (Rev. ed.). New York: Oxford University Press.

Mooney, H. F. (1968). "Popular Music Since the 1920s: The Significance of Shifting Taste," *American Quarterly, 20*, pp. 67–85.

Morreale, Joanne. (1991). *A New Beginning: A Textual Frame Analysis of the Political Campaign Film.* Albany: State University of New York Press.

Morris, Desmond. (1977). *Manwatching: A Field Guide to Human Behavior.* New York: Abrams.

Muth, R. Timothy. (1991). "Hypertext Can Speed Research," *National Law Journal, 14,* pp. 39–41ff.

Narasimhan, R. (1991). "Literacy: Its Characterization and Implications." In David R. Olson and Nancy Torrance, Eds., *Literacy and Orality* (pp. 177–197). New York: Cambridge University Press.

Nichols, Bill. (1981). *Ideology and the Image: Social Representation in the Cinema and Other Media.* Bloomington: Indiana University Press.

Nielsen Report. (1991, July 8). In Bill Carter "TV Industry Unfazed by Rise in 'Zapping.' " *New York Times,* pp. D1, D6.

Nilan, Michael S. (1992, Autumn). "Cognitive Space: Using Virtual Reality for Large Information Resource Mangagement Problems," *Journal of Communication, 42,* pp. 115–135.

Nilles, Jack M. (1986). "Future Impacts of Information Technologies." In *1987 Yearbook of Science and the Future* (pp. 513–520). Chicago: Encyclopaedia Britannica.

Noble, Barbara Presley. (1994, June 15). "Electronic Liberation or Entrapment?" *New York Times,* p. C4.

Noble, David W. (1965). *Historians Against History: The Frontier Thesis and the National Covenant in American Historical Writing Since 1830.* Minneapolis: University of Minnesota Press.

Oliver, Robert T., and Marvin G. Bauer, Ed. (1959). *Re-Establishing the Speech Profession: The First Fifty Years.* New York: Speech Association of the Eastern States.

Olson, David R. (1974). "Introduction." In David R. Olson, Ed., *Media and Symbols: The Forms of Expression, Communication, and Education* (pp. 1–24). Chicago: University of Chicago Press.

Olson, David R. (1977). "Oral and Written Language and the Cognitive Processes of Children," *Journal of Communication, 27,* pp. 10–26.

Olson, David R. (1980). "On the Language and Authority of Textbooks," *Journal of Communication, 30,* pp. 186–196.

Olson, David R. (1986). "The Cognitive Consequences of Literacy," *Canadian Psychology, 27,* pp. 109–121.

Olson, David R. (1987). "An Introduction to Understanding Literacy," *Interchange, 18,* pp. 1–8.

Olson, David R. (1988). "Mind and Media: The Epistemic Functions of Literacy," *Journal of Communication, 38,* pp. 27–36.

Olson, David R. (1991a). "Literacy and Objectivity: The Rise of Modern Science." In David R. Olson and Nancy Torrance, Eds., *Literacy and Orality* (pp. 149–164). New York: Cambridge University Press.

Olson, David R. (1991b). "Literacy as Metalinguistic Activity." In David R. Olson and Nancy Torrance, Eds., *Literacy and Orality* (pp. 251–270). New York: Cambridge University Press.

Olson, David R., and Jerome S. Bruner. (1974). "Learning Through Experience and Learning Through Media." In David R. Olson, Ed., *Media and Symbols: The Forms of Expression, Communication, and Education* (pp. 125–150). Chicago: National Society for the Study of Education.

Olson, David R., and Nancy Torrance. (1987). "Language, Literacy, and Mental States," *Discourse Processes, 10,* pp. 157–167.

Olson, David R., and Nancy Torrance, Eds. (1991). *Literacy and Orality.* New York: Cambridge University Press.

Olson, Scott R. (1987). "Meta-television: Popular Postmodernism," *Critical Studies in Mass Communication, 4,* pp. 284–300.

O'Neill, James M. (1923). "Speech Content and Course Content," *Quarterly Journal of Speech Education, 9,* pp. 25–52.

Ong, Walter J. (1967). *The Presence of the Word: Some Prolegomena for Cultural and Religious History.* Minneapolis: University of Minnesota Press.

Ong, Walter J. (1971). *Rhetoric, Romance, and Technology: Studies in the Interaction of Expression and Culture.* Ithaca, NY: Cornell University Press.

Ong, Walter J. (1977). *Interfaces of the Word: Studies in the Evolution of Consciousness and Culture.* Ithaca, NY: Cornell University Press.

Ong, Walter J. (1980). "Literacy and Orality in Our Times," *Journal of Communication, 30,* pp. 197–204.

Ong, Walter J. (1982). *Orality and Literacy: The Technologizing of the Word.* New York: Methuen.

Opland, Jeff. (1983). *Xhosa Oral Poetry: Aspects of a Black South African Tradition.* Cambridge, UK: Cambridge University Press.

Parry, Adam, Ed. (1987). *The Making of Homeric Verse: The Collected Papers of Milman Parry.* New York: Oxford University Press.

Pellegrini, A. D., Lee Galda, Janna Dresden, and Susan Cox. (1991, May). "A Longitudinal Study of the Predictive Relations Among Symbolic Play, Linguistic Verbs, and Early Literacy," *Research in the Teaching of English, 25,* pp. 219–235.

Perez, Ernest. (1992, December). "The Linear File: Hypertext Is Growing Up." *Database,* pp. 8–9ff.

Perkinson, Henry J. (1995). *How Things Got Better: Speech, Writing, Printing, and Cultural Change.* Westport, CT: Bergin & Garvey.

Peterson, Eric E. (1987). "Media Consumption and Girls Who Want to Have Fun," *Critical Studies in Mass Communication, 4,* pp. 37–50.

Pfaffenberger, Bryan. (1992). "Technological Dramas," *Science, Technology, and Human Values, 17,* pp. 282–312.

Pfau, Michael. (1990, Spring). "A Channel Approach to Television Influence," *Journal of Broadcasting and Electronic Media, 34,* pp. 195–214.

Phelps, M. E., J. C. Mazziotta, and S. C. Huang. (1982). "Review: Study of Cerebral Function with Positron Computer Tomography," *Journal of Cerebral Blood Flow and Metabolism, 2,* pp. 113–162.

Phillips, Gerald M. (1994). "Introduction to the 'Internet,'" *Communication Education, 43,* pp. 71–72.

Phillips, Gerald M., and Bradley R. Erlwein. (1988). "Composition on the

Computer: Simple Systems and Artificial Intelligence," *Communication Quarterly*, 36, pp. 243–261.

Phillips, Gerald M., and Julia T. Wood, Eds. (1990). *Speech Communication: Essays to Commemorate the 75th Anniversary of the Speech Communication Association*. Carbondale: Southern Illinois University Press.

Pickens, Donald K. (1981). "The Historical Images in Republican Campaign Songs," *Journal of Popular Culture*, 15, pp. 165–174.

Porat, Marc Uri. (1977, May). *The Information Economy: Definition and Measurement* (Vol. 1). Washington, DC: Office of Telecommunications of the U.S. Department of Commerce.

Porat, Marc Uri, with the assistance of Michael Rogers Rubin. (1977, May). *The Information Economy: National Income, Workforce, and Input–Output Accounts* (Vol. 8). Washington, DC: Office of Telecommunications of the U.S. Department of Commerce.

Porter, R. H., J. M. Cernoch, and R. D. Balogh. (1985). "Odor Signatures and Kin Recognition," *Physiology and Behavior, 34*, pp. 445–448.

Porter, R. H., J. M. Cernoch, and F. J. McLaughlin. (1983). "Maternal Recognition of Neonates Through Olfactory Cues," *Physiology and Behavior, 30*, pp. 151–154.

Porter, R. H., and J. D. Moore. (1981). "Human Kin Recognition by Olfactory Cues," *Physiology and Behavior, 27*, pp. 493–495.

Postman, Neil. (1985). *Amusing Ourselves to Death: Public Discourse in the Age of Show Business*. New York: Viking Penguin.

Potter, W. James. (1988). "Perceived Reality in Television Effects Research," *Journal of Broadcasting and Electronic Media, 32*, pp. 23–41.

Poulakos, John. (1990). "Hegel's Reception of the Sophists," *Western Journal of Speech Communication, 54*, 160–171.

Poulakos, Takis. (1990). "Historiographies of the Tradition of Rhetoric: A Brief History of Classical Funeral Orations," *Western Journal of Speech Communication, 54*, pp. 172–188.

Press, Andrea L. (1991a). *Women Watching Television: Gender, Class, and Generation in the American Television Experience*. Philadelphia: University of Pennsylvania Press.

Press, Andrea L. (1991b). "Working-Class Women in a Middle-Class World: The Impact of Television on Modes of Reasoning about Abortion," *Critical Studies in Mass Communication, 8*, pp. 421–441.

Quart, Leonard, and Albert Auster. (1984). *American Film and Society Since 1945*. New York: Praeger.

Raffman, Diana. (1992). "Proposal for a Musical Semantics." In Mari Riess Jones and Susan Holleran, Eds., *Cognitive Bases of Musical Communication* (pp. 23–31). Washington, DC: American Psychological Association.

Rasmussen, Karen. (1994). "Transcendence in Leonard Bernstein's *Kaddish Symphony*," *Quarterly Journal of Speech, 80*, pp. 150–173.

Reagon, Bernice Johnson. (1975). *Songs of the Civil Rights Movement 1955–1965: A Study in Culture History*. Ann Arbor, MI: University Microfilms.

Regian, J. Wesley, Wayne L. Shebilske, and John M. Monk. (1992, Autumn).

"Virtual Reality: An Instructional Medium for Visual-Spatial Tasks," *Journal of Communication, 42*, pp. 136–149.

Reid, Loren. (1978). *Hurry Home Wednesday: Growing Up in a Small Missouri Town, 1905–1921*. Columbia: University of Missouri Press.

Reid, Loren. (1981). *Finally, It's Friday: School and Work in Mid-America, 1921–1933*. Columbia: University of Missouri Press.

Reid, Loren. (1990). *Speech Teacher: A Random Narrative*. Annandale, VA: Speech Communication Association.

Rein, Irving, J., and Craig M. Springer. (1986). "Where's the Music? The Problems of Lyric Analysis," *Critical Studies in Mass Communication, 3*, pp. 252–256.

Rheingold, Howard. (1991). *Virtual Reality*. New York: Simon & Schuster.

Ringer, R. Jeffrey, Ed. (1994). *Queer Words, Queer Images: Communication and the Construction of Homosexuality*. New York: New York University Press.

Robins, Robert Henry. (1987). "Language." In *The New Encyclopaedia Britannica* (Vol. 22, pp. 566–589). Chicago: Encyclopaedia Britannica.

Rogers, Everett M. (1986). *Communication Technology: The New Media in Society*. New York: Free Press.

Rogers, Everett M. (1994). *A History of Communication Study: A Biographical Approach*. New York: Free Press.

Roland, Alex. (1992, Winter). "Theories and Models of Technological Change: Semantics and Substance," *Science, Technology, and Human Values, 17*, pp. 79–100.

Roszak, Theodore. (1986). *The Cult of Information: The Folklore of Computers and the True Art of Thinking*. New York: Pantheon Books.

Roth, Lane. (1981). "Folk Songs as Communication in John Ford's Films," *Southern Speech Communication Journal, 46*, 390–396.

Rothenbuhler, W. E. (1981). *The Popular Music Industry: Financial Stability Versus Financial Success*. Paper presented at the meeting of the 1981 Conference on Culture and Communication, Philadelphia.

Rowland, Lucy M. (1994). "Libraries and Librarians on the Internet," *Communication Education, 43*, pp. 143–150.

Rubin, Alan M. (1985). "Uses of Daytime Television Soap Operas by College Students," *Journal of Broadcasting and Electronic Media, 29*, pp. 241–258.

Rubin, Alan M. (1993, March). "Audience Activity and Media Use," *Communication Monographs, 60*, pp. 98–105.

Rubin, Michael Rogers, Mary Taylor Huber, and Elizabeth Lloyd Taylor. (1986). *The Knowledge Industry in the United States 1960–1980*. Princeton, NJ: Princeton University Press.

Ryan, Halford Ross. (1983). *American Rhetoric from Roosevelt to Reagan: A Collection of Speeches and Critical Essays*. Prospect Heights, IL: Waveland Press.

Ryan, Susan M. (1994). "Uncle Sam Online: Government Information on the Internet," *Communication Education, 43*, pp. 151–158.

Salamon, Julie. (1992, June 25). "Film: New Villains Threaten Gotham," *The Wall Street Journal*, p. A14.

Salomon, Gavriel. (1979). "Shape, Not Only Content: How Media Symbols Partake in the Development of Abilities." In Ellen Wartella, Ed., *Children*

Communicating: Media and the Development of Thought, Speech, Understanding (pp. 53–82). Beverly Hills, CA: Sage.

Salt, Barry. (1985). "Statistical Style Analysis of Motion Pictures." In Bill Nichols, Ed., *Movies and Methods* (Vol. 2, pp. 691–701). Berkeley, CA: University of California Press.

Sammet, Jean E. (1993). "Key High-Level Languages." In Anthony Ralston and Edwin D. Reilly, Eds., *Encyclopedia of Computer Science* (3rd ed., pp. 1471–1472). New York: Van Nostrand Reinhold.

Sanders, Barry. (1991). "Lie it as it Plays: Chaucer Becomes an Author." In David R. Olson and Nancy Torrance, Eds., *Literacy and Orality* (pp. 111–128). New York: Cambridge University Press.

Sanders, Barry. (1994). *A Is for Ox: Violence, Electronic Media, and the Silencing of the Written Word.* New York: Pantheon Books.

Sanders, Donald H. (1985). *Computers Today* (2nd ed.). New York: McGraw-Hill.

Sandro, Paul. (1985). "Signification in the Cinema." In Bill Nichols, Ed., *Movies and Methods* (Vol. 2, pp. 391–407). Berkeley: University of California Press.

Sanford, W.P. (1922). "The Problems of Speech Content," *Quarterly Journal of Speech Education, 8*, pp. 364–371.

Santoro, Gerald M. (1994). "The Internet: An Overview," *Communication Education, 43*, pp. 73–86.

Schaefer, Richard J. (1991). "Public Television Constituencies: A Study in Media Aesthetics and Intentions," *Journal of Film and Video, 43*, pp. 46–68.

Scheflen, Albert E. (1972). *Body Language and Social Order: Communication as Behavioral Control.* Englewood Cliffs, NJ: Prentice-Hall.

Schlater, Robert. (1970). "Effect of Speed of Presentation on Recall of Television Messages," *Journal of Broadcasting, 14*, pp. 207–214.

Schoening, Gerald T., and James A. Anderson. (1995, May). "Social Action Media Studies: Foundational Arguments and Common Premises," *Communication Theory, 5*, pp. 93–116.

Scholes, Robert J., and Brenda J. Willis. (1991). "Linguists, Literacy, and the Intensionality of Marshall McLuhan's Western Man." In David R. Olson and Nancy Torrance, Eds., *Literacy and Orality* (pp. 215–235). New York: Cambridge University Press.

Scodari, Christine, and Judith Thorpe. (1992). *Media Criticism: Journeys in Interpretation.* Dubuque, IA: Kendall/Hunt.

Scott, Robert L. (1967). "On Viewing Rhetoric as Epistemic," *Central States Speech Journal, 18*, pp. 9–16.

Scott, Robert. (1970, September). "Rhetoric, Black Power, and Baldwin's 'Another Country,'" *Journal of Black Studies, 1*, pp. 21–34.

Scott, Robert L. (1976). "On Viewing Rhetoric as Epistemic: Ten Years Later," *Central States Speech Journal, 27*, pp. 258–266.

Scott, Robert L., and Wayne Brockriede. (1969). *The Rhetoric of Black Power.* New York: Harper & Row.

Scribner, Sylvia, and Michael Cole. (1981). *The Psychology of Literacy.* Cambridge, MA: Harvard University Press.

Shapiro, Michael A., and Annie Lang. (1991). "Making Television Reality:

Unconscious Processes in the Construction of Social Reality," *Communication Research, 18*, pp. 685–705.

Shapiro, Michael A., and Daniel G. McDonald. (1992, Autumn). "'I'm Not a Real Doctor, but I Play One in Virtual Reality: Implications of Virtual Reality for Judgments about Reality," *Journal of Communication, 42*, pp. 94–114.

Shirane, Haruo. (1990). "Lyricism and Intertextuality: An Approach to Shunzei's Poetics," *Harvard Journal of Asiatic Studies, 50*, pp. 71–85.

Sholle, David. (1991). "Reading the Audience, Reading Resistance: Prospects and Problems," *Journal of Film and Video, 43*, pp. 80–89.

Sholle, David. (1995, May). "No Respect! Disciplinarity and Media Studies in Communication. Resisting Disciplines: Repositioning Media Studies in the University," *Communication Theory, 5*, pp. 130–143.

Short, Kathy Gnagey. (1986). "Literacy as a Collaborative Experience: The Role of Intertextuality." In Jerome A. Niles, Ed., *National Reading Conference Yearbook* (Vol. 35, pp. 227–232).

Shotter, John, and Kenneth J. Gergen. (1994). "Social Construction: Knowledge, Self, Others, and Continuing the Conversation." In Stanley A. Deetz, Ed., *Communication Yearbook* (Vol. 17, pp. 3–33). Thousand Oaks, CA: Sage.

Shrum, L. J. (1995, August). "Assessing the Social Influence of Television: A Social Cognition Perspective on Cultivation Effects," *Communication Research, 22*, pp. 402–429.

Sloan, William David. (1991). *Perspectives on Mass Communication History*. Hillsdale, NJ: Erlbaum.

Smith, Arthur L., and Stephen Robb, Eds. (1971). *The Voices of Black Rhetoric: Selections*. Boston: Allyn & Bacon.

Smith, Robert Rutherford. (1980). *Beyond the Wasteland: The Criticism of Broadcasting*. Annandale, VA: Speech Communication Association; and Urbana, IL: ERIC Clearinghouse on Reading and Communication Skills of the National Institute of Education.

Smith, Stephen A. (1980). "Songs of the South: The Rhetorical Saga of Country Music Lyrics," *Southern Speech Communication Journal, 45*, 164–172.

Smith, Stephen A. (1994). "Communication and the Constitution in Cyberspace," *Communication Education, 43*, pp. 87–101.

Snow, C.P. (1962). *Two Cultures and the Scientific Revolution*. New York: Cambridge University Press.

Snow, Robert P. (1983). *Creating Media Culture*. Beverly Hills, CA: Sage.

Sommer, Robert. (1969). *Personal Space: The Behavioral Basis of Design*. Englewood Cliffs, NJ: Prentice-Hall.

Speech Communication Association. (1964). *Golden Anniversary*. New York: Author.

Springer, Sally P., and Georg Deutsch. (1981). *Left Brain, Right Brain*. San Francisco: Freeman.

Staats, Arthur W. (1968). *Learning, Language, and Cognition: Theory, Research, and Method for the Study of Human Behavior and Its Development*. New York: Holt, Rinehart & Winston.

Sterling, Christopher H., and John M. Kittross. (1978). *Stay Tuned: A Concise History of American Broadcasting*. Belmont, CA: Wadsworth.

Steuer, Jonathan. (1992, Autumn). "Defining Virtual Reality: Dimensions Determining Telepresence," *Journal of Communication, 42*, pp. 73–93.

Stevens, John D., and Hazel Dicken Garcia. (1980). *Communication History*. Beverly Hills, CA: Sage.

Stock, Brian. (1983). *The Implications of Literacy: Written Language and Models of Interpretation in the Eleventh and Twelfth Centuries*. Princeton, NJ: Princeton University Press.

Stone, Michael K. (1976). "Heav'n Rescued Land: American Hymns and American Destiny," *Journal of Popular Culture, 10*, pp. 133–141.

Stratton, Valerie N., and Annette Zalanowski. (1984). "The Effect of Background Music on Verbal Interaction Groups," *Journal of Music Therapy, 21*, pp. 16–26.

Street, Brian V. (1984). *Literacy in Theory and Practice*. New York: Cambridge University Press.

Streeter, Thomas. (1995, May). "No Respect! Disciplinarity and Media Studies in Communication. Introduction: For the Study of Communication and Against the Discipline of Communication," *Communication Theory, 5*, pp. 117–129.

Swanson, David L. (1972). "The New Politics Meets the Old Rhetoric: New Directions in Campaign Communication Research," *Quarterly Journal of Speech, 58*, pp. 31–40.

Taylor, Bryan C. (1993). "*Fat Man and Little Boy*: The Cinematic Representation of Interests in the Nuclear Weapons Organization," *Critical Studies in Mass Communication, 10*, pp. 367–394.

"Technology." (1987). In *The New Encyclopaedia Britannica* (Vol. 11, p. 601). Chicago: Encyclopaedia Britannica.

Thaut, Michael H. (1990a). "Neuropsychological Processes in Music Perception and Their Relevance in Music Therapy." In Robert F. Unkefer, Ed., *Music Therapy in the Treatment of Adults with Mental Disorders: Theoretical Bases and Clinical Interventions* (pp. 3–32). New York: Schirmer Books.

Thaut, Michael H. (1990b). "Physiological and Motor Responses to Music Stimuli." In Robert F. Unkefer, Ed., *Music Therapy in the Treatment of Adults with Mental Disorders: Theoretical Bases and Clinical Interventions* (pp. 33–49). New York: Schirmer Books.

Thayer, Lee, Ed. (1980). *Ethics, Morality and the Media: Reflections on American Culture*. New York: Hastings House.

Thomas, Cheryl Irwin. (1974). " 'Look What They've Done To My Song, Ma': The Persuasiveness of Song," *Southern Speech Communication Journal, 27*, pp. 260–268.

Thomas, Sari. (1986). "Mass Media and the Social Order." In Gary Gumpert and Robert Cathcart, Eds., *Inter/Media: Interpersonal Communication in a Media World* (3rd ed., pp. 611–627). New York: Oxford University Press.

Thompson, Robert L. (1947). *Wiring a Continent: The History of the Telegraph Industry in the United States, 1832–1866*. Princeton, NJ: Princeton University Press.

Thonssen, Lester, and A. Craig Baird. (1948). *Speech Criticism: The Development of Standards for Rhetorical Appraisal*. New York: Ronald Press.

Thonssen, Lester, A. Craig Baird, and Waldo W. Braden. (1970). *Speech Criticism* (2nd ed.). New York: Ronald Press.

Tiemens, Robert K. (1970). "Some Relationships of Camera Angle to Communicator Credibility," *Journal of Broadcasting, 14*, pp. 483–490.

Toffler, Alvin. (1990). *Powershift: Knowledge, Wealth, and Violence at the Edge of the 21st Century*. New York: Bantam.

Toyama, Jean Yamasaki. (1990). "Intertextuality and the Question of Origins: A Japanese Perspective," *Comparative Literature Studies, 27*, pp. 311–323.

Tubbs, Stewart L., and Sylvia Moss. (1994). *Human Communication* (7th ed.). New York: McGraw-Hill.

Turkle, Sherry. (1984). *The Second Self: Computers and the Human Spirit*. New York: Simon & Schuster.

Turkle, Sherry. (1995). *Life on the Screen: Identity in the Age of the Internet*. New York: Simon & Schuster.

Turow, Joseph. (1992). *Media Systems in Society: Understanding Industries, Strategies, and Power*. New York: Longman.

Unwin, Philip Soundy, and George Unwin. (1987). "Publishing." In *The New Encyclopaedia Britannica* (Vol. 26, pp. 457–492). Chicago: Encyclopaedia Britannica.

Vaughn, Dai. (1985). "The Space Between Shots." In Bill Nichols, Ed., *Movies and Methods* (Vol. 2, pp. 703–714). Berkeley: University of California Press.

Viera, Maria. (1990). "The Work of John Cassavetes: Script, Performance Style, and Improvisation," *Journal of Film and Video, 42*, pp. 34–40.

Vonnegut, Kristin S. (1992). "Listening for Women's Voices: Revisioning Courses in American Public Address," *Communication Education, 41*, pp. 26–39.

Wachtel, Edward. (1978). "Technological Cubism: The Presentation of Space and Time in Multi Image," *Etc., 4*, pp. 376–382.

Wallace, Karl R., Ed. (1954). *History of Speech Education in America*. New York: Appleton-Century-Crofts.

Walter, Otis, and Robert L. Scott. (1973). *Thinking and Speaking: A Guide to Intelligent Oral Communication*. New York: Macmillan.

Waltz, David L. (1982, October). "Artificial Intelligence," *Scientific American*, pp. 118–133.

Wander, Philip. (1983). "The Ideological Turn in Modern Criticism," *Central States Speech Journal, 34*, pp. 1–18.

Weaver, David H., and Richard G. Gray. (1980). "Journalism and Mass Communication Research in the United States: Past, Present, and Future." In G. Cleveland Wilhoit and Harold de Bock, Eds., *Mass Communication Review Yearbook* (Vol. 1, pp. 124–151). Beverly Hills, CA: Sage.

Weaver, Richard M. (1953). *The Ethics of Rhetoric*. Chicago: Regnery.

Webster, James G., and Lawrence W. Lichty. (1991). *Ratings Analysis: Theory and Practice*. Hillsdale, NJ: Erlbaum.

Webster's New Collegiate Dictionary. (1981). Springfield, MA: Merriam-Webster..

Webster's Third New International Dictionary of the English Language Unabridged and Seven Language Dictionary. (1986). Springfield, MA: Merriam-Webster.

Weisman, Eric Robert. (1985). "The Good Man Singing Well: Stevie Wonder as Noble Lover," *Critical Studies in Mass Communication, 2,* 136–151.

Wichelns, Herbert A. (1923). "Our Hidden Aims," *Quarterly Journal of Speech Education, 9,* pp. 315–323.

Wichelns, Herbert A. (1966). "The Literary Criticism of Oratory." In Donald C. Bryant, Ed., *The Rhetorical Idiom: Essays in Rhetoric, Oratory, Language, and Drama.* New York: Russell & Russell. (Original work published 1925)

Williams, Frederick. (1982). *The Communications Revolution.* Beverly Hills, CA: Sage.

Williams, Frederick. (1991). *The New Telecommunications: Infrastructure for the Information Age.* New York: Free Press.

Williams, Robert C. (1965). "On the Value of Varying TV Shots," *Journal of Broadcasting, 9,* pp. 33–43.

Wilson, C. Edward. (1974, Spring). "The Effect of Medium on Loss of Information," *Journalism Quarterly, 51,* pp. 111–115.

Winans, James. (1923). "Speech," *Quarterly Journal of Speech Education, 9,* 223–231.

Windahl, S. (1981). "Uses and Gratifications at the Crossroads." In G. Cleveland Wilhoit and Harold deBock, Eds., *Mass Communication: Review Yearbook* (Vol. 1, pp. 174–185). Beverly Hills, CA: Sage.

Windt, Theodore Otto, Jr. (1990). *Rhetoric as a Human Adventure: A Short Biography of Everett Lee Hunt.* Annandale, VA: Speech Communciation Association.

Winner, E. (1982). *Invented Worlds.* Cambridge, MA: Harvard University Press.

Winner, Langdon. (1980). "A Response to Stan Carpenter's Review of *Autonomous Technology,*" *Research in Philosophy and Technology, 3,* pp. 124–126.

Wood, Barbara S. (1976). *Children and Communication: Verbal and Nonverbal Language Development.* Englewood Cliffs, NJ: Prentice-Hall.

Woolgar, Steve. (1991). "The Turn to Technology in Social Studies of Science," *Science, Technology, and Human Values, 16,* pp. 20–50.

Work, William, and Robert C. Jeffrey. (1989). *The Past Is Prologue: A 75th Anniversary Publication of the Speech Communication Association.* Annandale, VA: Speech Communication Association.

Wurman, Saul. (1989). *Information Anxiety.* New York: Doubleday.

Wurtzel, Alan H. (1983). *Television Production* (2nd ed.). New York: McGraw-Hill.

Wurtzel, Alan H., and Joseph R. Dominick. (1971). "Evaluation of Television Drama: Interaction of Acting Styles and Shot Selection," *Journal of Broadcasting, 16,* pp. 103–110.

Yates, Frances A. (1966). *The Art of Memory.* Chicago: University of Chicago Press.

Zemlin, Willard R. (1981). *Speech and Hearing Science: Anatomy and Physiology* (2nd ed.). Englewood Cliffs, NJ: Prentice-Hall.

Zettle, Herbert. (1973). *Sight, Sound, Motion: Applied Media Aesthetics*. Belmont, CA: Wadsworth.

Zettle, Herbert. (1984). *Television Production Handbook* (4th ed.). Belmont, CA: Wadsworth.

Zuboff, Shoshana. (1988). *In the Age of the Smart Machine: The Future of Work and Power*. New York: Basic Books.

Zumthor, Paul. (1990). *Oral Poetry: An Introduction*. Minneapolis: University of Minnesota Press.

Index

About the Authors

James W. Chesebro (Ph.D., University of Minnesota) has been a professor in the Department of Communication at Indiana State University, Terre Haute, since 1992. In 1996, he also served as President of the national Speech Communication Association. From 1989 to 1992, he was the Director of Educational Services in the National Office of the Speech Communication Association in Annandale, Virginia. From 1986 through 1988, Dr. Chesebro served as Chair of the Speech Communication Association's Publications Board and as a member of the Speech Communication Association's Administrative Committee and Legislative Council. He was the Editor of *Communication Quarterly* from 1985 through 1987. He was President of the regional Eastern Communication Association, in 1982–1983. Recently published books have included the following: In 1993, he edited *Extensions of the Burkeian System*; in 1990, he coedited the third edition of *Methods of Rhetorical Criticism: A Twentieth-Century Perspective*; and in 1989 he coauthored *Computer-Mediated Communication: Human Relationships in a Computerized World*. His articles have appeared in virtually every national and regional communication journal, including the *Quarterly Journal of Speech*, *Communication Monographs*, *Communication Education*, *Text and Performance Quarterly*, and *Critical Studies in Mass Communication*. In 1985, Dr. Chesebro received the Speech Communication Association's "Golden Anniversary Award" for the outstanding monograph of the year. In 1989, he received the "Everett Lee Hunt Scholarship Award" and the "Distinguished Service Award" from the Eastern Communication Association. In 1990, he received the Speech Communication Association of Puerto Rico's "Jose De Diego Award for Outstanding Service to SCAPR and to the Hispanic Community." In 1993, he received the Kenneth Burke Society's Distinguished Service Award. In 1996, he was designated a Research Fellow of the Eastern Communication Association.

Dale A. Bertelsen (Ph.D., Pennsylvania State University) is a professor in the Department of Communication Studies at Bloomsburg University in Bloomsburg,

Pennsylvania. In 1996, he also served as President of the regional Eastern Communication Association. He has served as Editor of the *Kenneth Burke Society Newsletter* and as Editor of Publications of the national Kenneth Burke Society. In 1993, he received the Kenneth Burke Society's Emerging Scholar Award, and in 1996 he also received the Kenneth Burke Society's Distinguished Service Award. In addition to his work on Kenneth Burke, his published research explores the methods and perspectives of rhetorical criticism and has appeared in journals such as *Communication Education*, *Philosophy and Rhetoric*, and *Communication Quarterly*.